Being a One Percenter

A Life of Service

Roy Taylor

Being a One Percenter

Published by:

AUTHORS on MISSION

Table of Contents

Preface

Sirens wailed in the distance, gunshots echoed through the night, and a radio crackled with urgent messages. For most, these sounds are of danger and fear. For me, they've been the soundtrack of a life dedicated to service. However, in the quiet moments between the chaos of the emergency calls and the weight of life-altering decisions, I've often reflected on the journey that brought me here.

From a young boy in Amherst, Ohio, to a seasoned professional with a diverse career spanning law enforcement, military service, and academia, my path has been anything but ordinary. I've lived a life that most people only see in movies or read about in books. But this isn't fiction. It's a raw, unfiltered look at what it truly means to dedicate yourself to something greater than personal gain. This book, "Being a One Percenter: A Life of Service," is not just a recounting of my experiences but a testimony to the power of perseverance, lifelong learning, and unwavering dedication to serving others.

This isn't a book for the faint of heart. If you picked it up hoping for an easy path to success or a guaranteed formula for reaching the top, you should put it back on the shelf. But if you feel a burning desire to make a difference, stand on the front lines of society, and be part of the elite few who run towards danger when others flee – then keep reading. You might just be a "one percenter" in the making.

What's a one percenter, you ask? Forget what you've heard about wealthy elites or outlaw bikers. I'm talking about the rare individuals who dedicate their lives to protecting and serving others. The ones who wear uniforms not for show but as a badge of responsibility. These are the mere 1% of the world's population who serve in law enforcement and the military or earn advanced degrees to better understand and improve our world.

I've been all three, and let me tell you – it's not an easy road. It's a life sentence, but one I wouldn't trade for anything in the world.

But this book isn't about bragging rights or setting myself apart. It's about showing that extraordinary achievements are within reach for anyone willing to do the work, remain curious, and stay committed to personal growth. My journey proves that formal education isn't always the starting point for success. I began my career with modest academic achievements but a burning desire to make a difference. Through determination, adaptability, and a willingness to learn from every experience, I've crafted a life and career that has allowed me to impact countless lives positively.

As I write this preface, I realize that this book has been decades in the making. Every late-night patrol, every tense negotiation, every life-or-death decision in Iraq, and every challenging case as an expert witness has contributed to the story I'm about to share. The decision to finally put pen to paper (or fingers to keyboard, as it were) came from a deep-seated desire to pass on the lessons I've learned – both the triumphs and the missteps – to the next generation of public servants, leaders, and dreamers.

One particular moment stands out in my memory, crystalizing the need for this book. It was November 2016, and in a single week, my office received six cases from around the country, all involving police shootings of individuals in psychiatric crises. As an expert witness, I had encountered similar cases before, but never in such rapid succession. The sheer volume was staggering and pointed to a systemic issue that demanded attention.

This flood of cases catalyzed my doctoral research, which focused on crisis intervention training for law enforcement officers. It also highlighted a broader truth: our experiences, when properly analyzed and shared, can lead to meaningful change. This book is an extension of that belief—that by sharing our stories and the wisdom we've gained, we can help others navigate their paths more effectively.

But this isn't just my story. It's a roadmap for anyone who feels that same pull towards service I felt as a young man. Whether you're a young person considering a career in law enforcement or the military, a mid-career professional looking to make a more meaningful impact, or someone searching for inspiration to pursue your dreams, there's something here for you.

The world needs more people willing to answer the call to service. We need individuals who see challenges as opportunities, setbacks as learning experiences, and true success as being measured not by the size of your bank account but by your impact on others.

But let me be clear – this path isn't for everyone. It will test you in ways you can't imagine. There will be sleepless nights, missed family events, and moments of doubt that shake you to your core. You'll face danger, both physical and emotional. You'll make decisions that impact lives and always carry the weight of those choices.

If that sounds daunting, good. It should. This life isn't for the casual observer or the half-hearted participant. It's for those rare individuals willing to put it all on the line for something bigger than themselves.

Throughout these pages, you'll journey with me from my adoption as a young child to my rise through the ranks of law enforcement. You'll feel the heat of the Iraqi desert and the tension of high-stakes counterterrorism operations. You'll see how a lifelong commitment to learning and growth can take you from a struggling student to a Ph.D. holder and respected expert in your field.

My journey from a C-student in high school to earning a Ph.D. later in life proves that it's never too late to invest in your education and personal growth. I'll share the strategies I used to balance work, family, and academic pursuits, demonstrating that you can achieve more than you ever thought possible with the right mindset and approach.

Another key theme is the value of adaptability and seizing opportunities. Throughout my career, I've transitioned between various roles – from law enforcement to firefighting, from military service to private sector work, and even into academia and expert testimony. Each shift brought new challenges and opportunities for growth. I'll discuss how to recognize these moments of potential change and make the most of them.

We'll also delve into some of public service's more complex and challenging aspects, particularly law enforcement. I'll share my experiences and insights on critical issues such as mental health crisis response, the importance of community-oriented policing, and the ongoing need for police reform and accountability. These are difficult conversations, but they're essential for anyone looking to make a positive impact in this field.

This book isn't about glorifying service challenges—it's about preparing you to face them head-on and emerge stronger on the other side.

As you read, you'll meet others who have walked similar paths. From fellow officers who've risen through the ranks to soldiers who've faced the crucible of combat, these stories will show you that while the journey of service is often solitary, you're never truly alone.

For those with entrepreneurial aspirations, I'll recount the story of starting and growing my private police business, Capitol Special Police. You'll learn about the challenges of building a business from the ground up, the sacrifices required, and the rewards of creating something that serves a vital community need.

Throughout the book, I aim to balance sharing personal experiences and offering practical advice. You'll find actionable strategies for career advancement, tips for effective leadership, and insights into navigating complex organizational structures. But more importantly, you'll find inspiration to push yourself beyond your perceived limits and to always strive for excellence in the service of others.

By turning the final page, I hope you'll be inspired to reflect on your life's journey and your potential to make a difference. Whether you choose a path in public service, pursue higher education, start a business, or find your unique way to contribute to society, you'll have the tools and mindset to make it happen.

Remember, being a "one percenter" isn't about being better than others – it's about being willing to give, sacrifice, and push yourself further to serve the greater good.

So, are you ready to answer the call? Are you prepared to join the ranks of those who run toward danger, stand up for what's right, and dedicate their lives to serving others?

If so, turn the page. Your journey starts now.

And for those who decide this path isn't for them? That's okay, too. The world needs all kinds of people. But the next time you hear sirens in the night or see a uniform walking by, I hope you'll remember that there are men and women out there who have dedicated their lives to keeping you safe and making our world a little bit better.

This book is dedicated to those brave souls. To those considering joining their ranks, welcome to the adventure of a lifetime.

Chapter 1
Childhood in Adoption

As I sit here, reflecting on the journey that brought me to where I am today, I'm struck by how the smallest moments can shape a life's trajectory. My story begins not with my birth but with a decision made by two people who would become the cornerstones of my existence - my adoptive parents, William Fay Taylor and Ernestine Marie Taylor.

It was 1962 in Amherst, Ohio, a small town nestled about 30 miles west of Cleveland. Amherst was, and still is, one of those quintessential Midwest towns where everyone seems to know each other and where community isn't just a word but a way of life. It's the place that shapes you, molds your values, and instills a sense of belonging that stays with you no matter where life takes you.

I was just one year old when the Taylors decided to adopt me. My birth name was Troy, given by a young mother who, for reasons I may never fully understand, felt unable to handle the enormous responsibility of raising a child. It's a poignant reminder that sometimes, the most selfless act of love is letting go. I've come to respect her decision over the years, recognizing the courage it must have taken to put my needs above her desires.

Fay and Marie renamed me Roy - a small change that would come to symbolize the new life and identity they were offering me. Looking back, I realize that this act of renaming wasn't just about giving me a new moniker but a fresh start, a clean slate upon which to write my own story.

Let me pause here and share something I've learned through my years of service and leadership: our beginnings don't define us, but they do shape us. Whether you're born into adversity or advantage, what matters most is what you do with the hand you're dealt. My adoption could have been a story of loss, but instead, it became a foundation of love, support, and opportunity.

You might wonder why I'm starting my story here with events I can't remember. To understand where we're going, we must appreciate where we've come from. Each of us has a unique starting point in life, and recognizing that can help us better understand our journey and the journeys of those around us.

My mother, Marie, often told her friends that even as a toddler, I showed signs of the man I would become. "Ever since he was a little boy," she'd say, her voice tinged with pride and amusement, "he'd always been in uniform and always a leader." At the time, I probably just smiled at her gaze, not fully grasping the significance of what I was doing. But now, looking back on a career that's spanned law enforcement, military service, and beyond, I see the truth in her observation.

I remember countless times she'd tell me how I spent my afternoons organizing my model airplanes and toy soldiers into neat rows as if preparing them for inspection. My favorite games always involved some form of leadership or service - playing "Army" with my toy trucks or "GI Joe" dolls. These might seem like simple childhood games, but they were early expressions of the values that would guide my life: organization, leadership, and a desire to help others.

Here's a lesson I've carried with me throughout my life: the seeds of who we'll become are often planted early. Pay attention to what naturally draws your interest and gives you a sense of purpose, as these early inclinations can powerfully indicate your future path. My father, Fay, known to his friends and colleagues as "Workhorse," was a man of few words but decisive actions. Born in Alabama in 1933, he faced adversity early in life when his father passed away from cancer. He was just six months old, too young to form any memories of his dad. Years later, his mother remarried and moved to Ohio, blending with her new husband's family.

Dad rarely spoke about his childhood, but the impact of his early experiences was evident in his approach to life and work. He believed in the value of hard work, dedication, and providing for one's family - principles that would profoundly shape my worldview.

One of the most valuable lessons I learned from my father was the importance of resilience and adaptability. Life rarely goes according to plan, and our ability to adjust and keep moving forward in the face of challenges often determines our success. Whether it was adapting to a new family situation as a child or later juggling multiple jobs to provide for us, my father exemplified this resilience.

My mother, Marie, brought a different but equally important perspective to my upbringing. Originally from Tennessee, she had worked as a teacher in a one-room schoolhouse in the Appalachian Mountains. Her experiences there instilled in her a deep appreciation for education and the transformative power of learning.

I didn't always appreciate my mother's emphasis on education as a youth. Like many young people, I was more interested in action than academics. But her influence planted a seed that would grow throughout my life, culminating in my pursuit of a Ph.D. later in my career. It's a reminder that the value of education isn't always immediately apparent, but it's an investment that pays dividends throughout your life.

I never tired of hearing the story of how my parents met. It was a tale of chance encounters and quick romance that seemed almost like something out of a movie. One summer, my mother ventured north to Ohio with her cousin, seeking work at a golf ball factory. On a fateful Sunday morning, she attended a local church service.

As my mother recounted, her eyes twinkling with the memory, "I saw your father walking toward the front of the building with a tape recorder in his hand. He sat in the front row, and I thought he was the minister!"

I always chuckled at this part. The idea of my quiet, reserved father being mistaken for a minister seemed absurd. But as my mother explained, he was there to record the sermon for his disabled Aunt who couldn't attend church - a small act of kindness that spoke volumes about his character.

Beyond its romantic appeal, this story taught me an essential lesson about judgment and perception. How often do we make assumptions about people based on a first glance? My mother's initial impression of my father was utterly wrong, yet it led to a lifetime of love and partnership. It's a reminder to look beyond first impressions and to take the time to honestly know people before forming judgments.

My parents' paths frequently crossed in the following weeks, leading to a swiftly blossoming romance that culminated in their marriage in 1953. Shortly after, my father was drafted into the Army, serving his enlistment as a medic stationed at Fort Ritchie, Maryland.

His time in the service left a lasting impression on him and, by extension, on me. He often spoke of his days at Fort Ritchie, which serviced Camp David whenever the president was there. "There were many days," he would recount, "where I'd sit out on the flight line in an ambulance, waiting for the president's plane to land. We were there just in case there was some mishap." Even then, the idea of being ready to serve and help in crises fascinated me.

When Dad returned from the Army, he set his sights on working for the Ford Motor Company like many men in our area. They opened a new assembly plant in Lorain, and jobs were scarce. Dad would often recount how he got his job at Ford, a story that never failed to impress upon me the value of persistence and readiness.

“Every morning,” he'd tell me, “I'd go down to the plant with my lunch pail. Dozens of us would sit there in the parking lot, waiting. Then someone from HR would yell, 'I need five more!' The next five men would go in, get hired, and start their new jobs on the assembly line right then and there.”

Dad's persistence paid off, and he secured a job at Ford. But that was just the beginning of his work ethic, which earned him the nickname "Workhorse." For 32 years, he worked at that Ford plant. But for Dad, one job was never enough. In addition to his full-time work at Ford, he took on a part-time position as a police officer in our hometown of Amherst.

His schedule was grueling, and I was awed by his stamina as a child. He'd leave the house at 4:30 in the morning for his shift at the Ford plant. By 2:30 in the afternoon, he'd return home briefly to change into his police uniform before patrolling Amherst's streets from 3 to 11 p.m. But Dad's work didn't stop there. In addition to his regular police duties, he took on odd jobs around the department. He'd service the parking meters downtown, collecting money and fixing those that weren't working. Dad was there to help with any other task that needed to be done at the police department.

As a child, I didn't see much of my father. His work kept him away from home for long hours. But I found comfort in the rhythms of his comings and goings. Late at night, I'd listen for the distinctive clunk of the hallway light switch. That sound meant Dad was home safe, hanging his coat in the hallway closet. Only then could I truly relax and drift off to sleep.

Looking back, I realize that my father's police uniform profoundly impacted me. Seeing him come home nearly every night for dinner in that uniform planted a seed in my mind. Without even realizing it, I was drawn to the idea of a life in law enforcement, a life of service. Little wonder, when I was in second grade, I often found myself playing "cops and robbers" with my friends and reenacting scenes from police shows on television.

At that same time, like many boys my age, I was captivated by the space race and the allure of aviation. I dreamed of becoming an astronaut, venturing into the great unknown of outer space. I spent hours building model aircraft, progressing from simple U-control planes to more complex radio-controlled ones. Little did I know then how these seemingly disparate interests in law enforcement and aviation would converge later in my life.

While my father was the epitome of a hardworking provider, my mother, Marie, was the heart of our home. She embraced her role as a stay-at-home mother with the same dedication my father showed to his work. She was a typical example of a 1950s and '60s wife and child caregiver. Mom was a whirlwind of activity, ensuring our household ran like a well-oiled machine.

Every meal was home-cooked, and the laundry was always done—not just done, but to perfection. Mom would iron everything, from sheets to T-shirts and every garment we wore. I took this for granted as a child, but now I understand the love and care that went into every pressed collar and smooth pillowcase.

But Mom's activities extended far beyond our home. She was deeply involved in the PTA, attending meetings and volunteering for school projects. She made sure I always had the support I needed for my education, even if I didn't always appreciate it at the time.

One of my fondest memories is Mom reading to me every night before bed. Those nightly reading sessions improved my literacy and opened my mind to worlds beyond our small town.

As I continue my story, you'll see how these early influences – my father's work ethic and sense of duty, my mother's nurturing support and emphasis on education, and my innate drive to lead and serve – shaped my life. They laid the foundation for a lifelong commitment to serving others and being ready to help in times of need. This reminds us that our actions, even when we think they're insignificant, can profoundly impact others - especially our children.

Every experience, every person we encounter, shapes us in some way. The key is to remain open to these influences, learn from them, and use them as stepping stones to success and fulfillment. My journey from an adopted child in small-town Ohio to a life of service and leadership wasn't always straightforward. Still, these early experiences and influences planted the seeds for everything.

In the next part of our journey together, we'll explore how these early influences began manifesting in my childhood and teenage years, setting the stage for the path that would ultimately lead me to a life of service. Remember, your past doesn't determine your future, but understanding where you've come from can help you chart a course to where you want to go.

Chapter 2

The Boy Scout

As I settled into my new life growing up with the Taylors, I later learned that their commitment to service extended far beyond their decision to adopt me. Their actions, deeply rooted in their Christian faith, would shape my understanding of living a life of service.

Now, you might wonder, what exactly is a "life of service"? I've pondered this question many times over the years, and it's worth taking a moment to reflect on it.

As I've come to understand it, a life of service is a commitment to putting the needs of others before your own. It's about waking up each day to make the world better than you found it. It's not always grand gestures or heroic acts—more often than not, the small, consistent daily choices genuinely define a life of service.

It's the police officer who takes an extra moment to comfort a frightened child at the scene of an accident. It's the teacher who stays late to help a struggling student. A life of service is the neighbor who mows the lawn for the elderly couple next door without being asked. And yes, the parents open their hearts and homes to children needing love and stability.

You see, my parents, Fay, and Marie Taylor, weren't content with just adopting me. Their hearts were so full of love and compassion that they decided to open their home to another child in need. Several years after they welcomed me, they adopted my brother Nathan.

But before then, it all happened that one Sunday morning, after I had joined the family, we attended our local church service as usual. There, my parents noticed a family struggling to care for their children. The parents looked overwhelmed, their faces etched with worry and fatigue. Without hesitation, my parents approached them after the service, offering to help however they could.

This family, as it turned out, was facing significant challenges. They had a crossed-eye child among several other children and were finding it difficult to provide each child's care and attention. In an act of incredible generosity, my parents offered to unofficially adopt and take in one of their boys for a while to help ease their burden. It was that child they took.

They took Nathan and raised him as their own. In 1965, they decided to adopt another child, and my brother Nathan, named after Nathan, the quasi-foster child, came home. My brother's adoption further expanded our family and demonstrated their commitment to providing loving homes for children who needed them.

This act of opening their hearts and homes multiple times speaks volumes about Fay and Marie Taylor's character. It wasn't just about having children; it was about making a difference in the lives of those who needed it most.

Growing up in this household, I was surrounded by love, support, and strong moral values. My parents didn't just talk about the importance of helping others; they lived it daily through their actions. Right from an early age, they also tried instilling discipline and work ethic in us.

They made us do simple tasks like making our bed in the morning or helping with chores around the house. I sometimes felt frustrated watching my younger brother struggle with things that came naturally to me. This was years later while growing up. But at the time, my parents approached the situation with patience and understanding. They saw this not as a burden but an opportunity to make a real difference in a child's life.

Shortly after my younger brother Nathan was adopted, they returned the quasi-foster child to his family. Over the weeks and months he had stayed with us, my parents worked tirelessly to help this boy develop the values and habits that would serve him well. They taught him the importance of responsibility, hard work, and the satisfaction of a well-done job. It wasn't always easy, and there were moments of tension and difficulty. But gradually, they began to see growth in him.

By the time he returned to his biological parents, he had transformed. He had developed a strong work ethic, learned the importance of discipline, and gained a new self-respect. This experience taught me a powerful lesson about the impact of love, patience, and consistent guidance on a person's life.

My younger brother and I grew up together in the home, but our similarities ended there. My brother enjoyed sports, but I enjoyed playing "Army" in the woods or riding my bike. We were opposite in many ways, and it was clear that he lacked some of the discipline and work ethic that my parents had tried to instill in us from an early age.

I'd tackle any new assignments immediately and work to complete them as soon as possible. But Nathan, he'd wait until the last night before the assignment was due and then tell my mom about it. There were many long late nights spent finishing school assignments. For sure, Mom always comes through for him.

Looking back on this experience solidified my understanding of what it means to live a life of service. Living a life of service isn't always easy. It often requires sacrifice, patience, and perseverance. It means being willing to sacrifice comfort, time, and sometimes even safety for the benefit of others. But in my experience, it's also gratifying. A life of service is about recognizing our interconnectedness and understanding that we elevate ourselves and our entire community by lifting others.

A deep sense of fulfillment comes from knowing you've made a positive difference in someone else's life, no matter how small. As we go through life and make daily decisions, we must learn to put others' needs before our own and extend a helping hand even when it's inconvenient or challenging.

As I progressed through those early years, I was drawn to activities that involved helping others. Whether assisting the teacher in school in handing out materials or helping a classmate struggling with an assignment, I discovered a sense of satisfaction in being useful and making things a little easier for those around me.

This innate desire to help made scouting a natural fit for me when I joined the Cub Scouts at eight in the fall of 1970. I had just started second grade, and my parents, always looking for ways to enrich my life and instill good values, suggested I try it. At the time, I was more interested in playing with my friends and building aircraft, but something about the uniforms and the promise of adventure caught my attention.

I remember my first pack meeting. The smell of church basement coffee mingled with the crisp scent of new uniforms, the excited chatter of boys my age, and the proud smiles of parents watching from the sidelines. As I stood in my brand new blue uniform, fidgeting with the unfamiliar neckerchief, I had no idea I was taking the first steps on a lifelong journey of service as a man in uniform.

The Scouting movement, founded by Robert Baden-Powell in the early 20th century, has a rich history of nurturing young people to become responsible, caring, and capable adults. Its emphasis on outdoor skills, community service, and character development aligned perfectly with the values my parents had already instilled in me.

From the first den meeting, I was captivated by the sense of belonging and purpose that Scouting offered. We weren't just a group of boys playing games and earning badges; we were part of something bigger, a worldwide brotherhood dedicated to positively impacting our communities and the world.

The Scout Law, which I can still recite without hesitation, became a guiding force in my life:

"A Scout is trustworthy, loyal, helpful, friendly, courteous, kind, obedient, cheerful, thrifty, brave, clean, and reverent."

These weren't just words to memorize but principles to live by. Each week, as we gathered for our den meetings, we would focus on one aspect of the Scout Law, discussing what it meant and how to apply it in our daily lives.

I remember one particular meeting where we focused on being "helpful." Our den leader, a kind-faced man with salt-and-pepper hair and glasses that always seemed to slip down his nose, challenged us to find ways to help at home without being asked.

"Boys," he said, leaning in conspiratorially, "I want you to surprise your parents this week. Do a chore they usually have to remind you to do, but do it before they ask. Then, at our next meeting, we'll share how it went."

That week, I made my bed every morning without prompting and volunteered to help with the dishes after dinner. The look of pleasant surprise on my mother's face was worth far more than any merit badge. It was my first real taste of how good it feels to serve others, to lighten someone else's load.

As I progressed through the ranks of Cub Scouts, from Bobcat to Wolf, Bear, and finally Webelos, I found myself more drawn to the leadership aspects of scouting. I loved the camaraderie of our pack, the sense of belonging to something bigger than myself. But more than that, I loved the opportunities to lead.

By the time I reached Webelos, I had earned the "Arrow of Light," the highest rank in Cub Scouts; I was often asked to lead our den in activities or help younger scouts with their achievements. Here, I discovered my knack for organizing and motivating others, skills that would serve me well throughout my law enforcement and military career.

But scouting wasn't just about leadership and service. It was also about discipline, integrity, and perseverance. We learned to wake up early for camping trips, to push through when hikes got tough, and always to leave a campsite cleaner than we found it. These lessons in self-discipline and responsibility were invaluable.

I remember one particularly challenging camping trip when I was about ten years old. We had hiked several miles into the woods, each carrying our gear. As we set up camp, dark clouds began to gather, and soon, we were caught in a downpour. Some of the younger scouts were ready to give up, to call for our parents to come and get us, but our Scoutmaster gathered us around the sputtering campfire.

"Boys," he said, his voice barely audible over the rain, "this is where we learn what we're made of. We can either let this rain defeat us or rise to the challenge. What do you say we show Mother Nature what scouts are capable of?"

His words ignited something in us. We rallied together, rigging up extra tarps, digging drainage trenches around our tents, and even getting the fire strong enough to cook dinner. That night, as we huddled in our damp sleeping bags, I felt a sense of pride and accomplishment I had never experienced before. We had faced adversity and come out stronger.

This experience taught me a valuable lesson about resilience and teamwork, one that would serve me well in my future career. Whether facing down criminals or navigating the complexities of military operations, I often think back to that rainy night in the woods and draw strength from the memory.

As I continued in scouting, eventually transitioning to Boy Scouts, the lessons in leadership, service, and character development only deepened. We took on more challenging projects, engaged in community service, and pushed ourselves to new limits. Each merit badge earned was not just a patch on a sash but a new skill acquired, a new horizon explored.

One project that stands out in my memory was when our troop decided to renovate a playground at a local elementary school. The equipment was old and rusted, and the ground was uneven and full of weeds. Throughout several weekends, we worked tirelessly—sanding and painting the swing set, laying new wood chips, planting flowers, and even building a new sandbox.

The day we finished, we invited the schoolchildren to see their new playground. The joy on their faces as they ran to the swings and slides was unforgettable. In that moment, I understood one of the true powers of service—the ability to bring happiness to others through our actions.

This project also taught me valuable lessons about project management, teamwork, and the importance of seeing a task through to completion. Little did I know then how often I would draw on these skills years later, whether organizing police operations or leading military units.

But Scouting didn't just prepare me for a career in public service. It instilled in me a lifelong love of learning and self-improvement. The badge system rewarded us for mastering new skills and knowledge and taught me the value of continuous growth and development. This mindset was one of the critical things that would later drive me to pursue advanced degrees and constantly seek new challenges and growth opportunities.

Unfortunately, as I progressed through the ranks of scouting, my goal to earn my Eagle Scout was short-lived due to a lack of parental involvement from other Scouts' parents. Mr. Vegahoff, our junior high school counselor, eventually grew tired of being the only adult to structure the activities and lead the organization. He ended up quitting.

One of the most critical lessons scouting taught me in those early years of my development was the value of integrity. The Scout Law's first point - "A Scout is trustworthy" - became a cornerstone of my personal and professional ethics. I learned early on that authentic leadership is built on trust, and trust is earned through consistent, honest actions.

I recall a situation during a camping trip when I accidentally broke another scout's lensatic compass. My first instinct was to hide the evidence to avoid getting in trouble, but the words of the Scout Law echoed in my mind. With a knot in my stomach, I approached the scout and our troop leader, admitted what I had done, and offered to replace the compass.

To my surprise, instead of punishment, I was met with understanding and appreciation for my honesty. This experience taught me a valuable lesson about the power of integrity, one that would guide me throughout my career in law enforcement. Whether testifying in court, reporting on an incident, or leading a team, I always strived to embody that first point of the Scout Law - to be trustworthy above all else.

Looking back on my childhood years, I'm struck by how formative they shaped the person I would become. The values instilled in me by my adoptive parents - love, compassion, and the importance of helping others - were reinforced and expanded through my experiences in scouting. The skills I learned - from basic knowledge to survival techniques, from project management to team leadership - would prove invaluable in my future roles.

But more than the practical skills or values, my childhood experiences instilled a sense of purpose. From a young age, they showed me the satisfaction of serving others. This sense of purpose would guide me through the challenges and triumphs of my career, constantly pushing me to strive for excellence in serving others.

As we progress in this narrative, you'll see how these early lessons laid the groundwork for my future endeavors. The path from adopted child to aiming to be an Eagle Scout was, in many ways, the first leg of my journey towards a life of service - a journey that would eventually lead me through the ranks of law enforcement, the crucible of military service, and beyond.

In the chapters, we'll explore how these foundational experiences intertwined with my academic journey, early forays into law enforcement, and eventual military service. But for now, I hope you'll take a moment to reflect on your own formative experiences. What early influences shaped your path? What values were instilled in you as a child continue to guide you today?

As you go through life, remember that a life of service doesn't happen overnight. It's built daily, choice by choice, through small acts of kindness, commitment to personal growth, and a willingness to put others before ourselves.

Remember, a life of service doesn't necessarily mean wearing a uniform or holding a public office. It's about approaching each day with the question, "How can I make a positive difference?" Whether you're a parent, a teacher, a business owner, or any other role, there are always opportunities to serve, lead, and make your corner of the world a little bit better.

As we continue this journey together, I encourage you to consider incorporating the spirit of service into your life. After all, in the words of the Scout slogan, the key is to "Do a Good Turn Daily." It's these small, consistent acts of service that, over time, can transform not just our own lives but the world around us.

Chapter 3

Struggles of My Early Education

Growing up in those earlier days, my first real taste of the world beyond our loving home came when I started school. For some unknown reason, this memory stayed with me. I was five years old, standing in the parking lot of what seemed like an impossibly large building. The air was filled with a mix of excitement and apprehension—not just mine but that of dozens of other children beginning their educational journey.

My parents, Fay and Marie Taylor, had just dropped me off. I remember the sound of our aqua-green Chevy Impala's engine fading as they drove away, leaving me to face this new adventure alone. It might seem cruel by today's standards, but back then, parents believed in fostering independence from an early age. Looking back, I'm grateful for that push.

There I stood, a small boy in a sea of unfamiliar faces, tasked with finding my way to a classroom I'd never seen before. The enormity of the school building loomed over me, and its doors seemed impossibly high. I recall reaching far above my head to grasp the large doorknob.

As I struggled with the door, I caught sight of something that would stick in my memory for years to come – a black ambulance parked nearby. At the time, it seemed like just another detail in a morning full of new experiences. Little did I know that this ambulance belonged to my soon-to-be teacher, Mrs. Majoris, whose husband ran a limousine and ambulance service in nearby Elyria.

It's funny how life comes full circle. Nearly two decades later, I worked for the Majoris family as an emergency medical technician – one of my first jobs after being discharged from the Air Force. Mrs. Majoris never tired of telling my coworkers how I had a crush on her in kindergarten. Life has a way of connecting the dots in the most unexpected ways.

But on that morning in 1967, all of that was in the future. At that moment, I was just a small boy facing his first real test of independence. As I finally managed to open the heavy door and step inside the school, I felt a mix of fear and determination. I had no idea where to go, but I knew I had to figure it out.

Looking back on that day now, I realize it was one of my first experiences with a challenge that would define much of my life – navigating unfamiliar territory and finding my way forward, even when the path wasn't clear.

You see, life rarely gives us a roadmap. Whether you're a five-year-old on your first day of school or a seasoned professional facing a new challenge, the feeling is often the same. You're in uncharted territory, and finding your way is up to you.

Over the years, I've learned that the key to success in these situations isn't having all the answers. It's about having the courage to take that first step, ask for help when needed, and keep moving forward even when unsure.

I found my classroom that day, though I couldn't tell you exactly how I managed it. I remember the sense of accomplishment I felt when I finally sat down at my little desk. It was a small victory but an important one. It taught me that I was capable of more than I thought.

Going through life, I have realized that every expert was once a beginner, and every leader once faced their first day in a new role. What separates those who succeed from those who don't isn't innate talent or luck – it's the willingness to step into the unknown, face challenges head-on, and learn and grow from every experience.

That first day of school was just the beginning of my many journeys and challenges. So, whether in your career, education, or personal life, the next time you face a new challenge, remember that little boy reaching for a doorknob that seemed just out of reach. Take a deep breath, summon your courage, and take that first step. You might be surprised at where it leads you.

But just because you overcame a challenge on your path to success doesn't mean others won't surface. As I look back on my journey from my early days in school to becoming an average student and then to earning a Ph.D., I'm reminded of a quote by Nelson Mandela: "Education is the most powerful weapon which you can use to change the world." But my path to embracing this truth was far from straightforward. I was the last person you'd expect to champion the value of formal education for much of my early life.

Picture a classroom in Amherst, Ohio, in the mid-1970s. The walls are adorned with colorful educational posters, and the air is filled with the scent of chalk dust and the rustle of paper. Seated near the back, a young boy with greased-down hair stares out the window, his mind elsewhere. That boy was me, and at the time, I couldn't have cared less about the lesson being taught.

It wasn't that I was incapable of learning. Instead, I couldn't see the point. My dream was to become a police officer, like my father, who worked part-time for the Amherst Police Department. In my young mind, I reasoned, "Why do I need to know algebra or the capital of Peru to chase bad guys?"

I'm somewhat embarrassed to admit that this attitude persisted throughout elementary and high school. I was what you'd call a solid C student—doing just enough to get by but never really applying myself. Honor rolls and academic accolades seemed like things for "other kids," not for me.

I vividly remember one parent-teacher conference in particular. Mrs. Cowan, my eighth-grade English teacher, sat across from my parents, her face a mixture of concern and frustration. "Roy has potential," she said, a phrase I'd heard countless times before, "but he just doesn't seem interested in applying himself."

My mother, Marie, nodded sympathetically as I squirmed in my chair, wanting to be anywhere but there. It wasn't that I didn't care about disappointing my parents – I did. But at that moment, I couldn't bridge the gap between their expectations and my lack of academic motivation.

Looking back now, I realize that my struggle wasn't unique. Many young people, perhaps even some of you reading this, have grappled with finding meaning and motivation in their studies. It's a challenge beyond simple laziness or rebellion – connecting what we learn in the classroom to our goals and aspirations in the real world.

I couldn't see the connection between algebra, literature, or world history and my dream of becoming a police officer. This mindset was not only shortsighted but potentially dangerous. In today's complex world, law enforcement requires a broad knowledge base and critical thinking skills that only a well-rounded education can provide.

Modern policing demands much more than physical fitness and knowledge of the law. Today, officers must understand complex social issues, communicate effectively with diverse communities, analyze data, and make split-second decisions that can have life-altering consequences. A solid educational foundation must be laid to develop these skills.

As I reflect on this journey, I'm reminded of another quote from Benjamin Franklin: "An investment in knowledge pays the best interest."

Education is an investment that pays dividends throughout one's life, and I encourage you to make it, regardless of where you are in your educational or professional journey.

If you struggle to find meaning in your studies or career, I urge you to look beyond the immediate tasks. Ask yourself: How does this connect to where I want to be in the next five or ten years? Sometimes, the connection isn't apparent, but it's almost always there.

My attitude towards education changed dramatically during my freshman year of high school, thanks to an unexpected encounter with my guidance counselor. I had become interested in the Air Force and harbored dreams of becoming a pilot. The Air Force Academy was the perfect path to achieve this goal.

Excitedly, I marched into the counselor's office and declared my intention to apply to the Academy. His response was a sobering wake-up call. "To get into one of the military academies," he explained, "you need to be on the honor roll and have a recommendation from a senator."

It was like a bucket of cold water poured on my dreams. I felt my heart sink, plunging into an abyss of despair. The vibrant colors of the bustling room around me faded into dull shades of gray. The honor roll seemed achievable with hard work, but obtaining a senator's recommendation seemed impossible. To my blue-collar mind, that sounded like something reserved for rich kids or political insiders. It was disheartening, and my chest tightened. Each beat resonated with an unsettling rhythm, a steady reminder of the crushing disappointment that wrapped around my thoughts.

But sometimes, a setback can be the catalyst for positive change. That conversation lit a fire under me. I realized that if I wanted to achieve my goals, I needed to apply myself academically. So, I buckled down and started taking my studies seriously for the first time.

By the end of my junior year, I had achieved a remarkable transformation by making it onto the honor roll and meeting all the requirements for early graduation. I felt pride and confusion when I was called into the office and informed of this. "What do I do now?" I asked.

The response I received could have been more inspiring. I was told I could enroll in college, enlist in the military, or return for my senior year to take woodshop and metal shop and have a senior study hall with early release. The idea of spending a year in what seemed like academic limbo felt like a colossal waste of time.

It was at this point that I began seeking alternatives. However, my search was hampered by a lack of knowledge – my own and, surprisingly, my guidance counselor's. I had no idea about the existence of private military academies like the Citadel or the Corps of Cadets at Virginia Tech. These institutions could have provided the perfect blend of military structure and academic rigor I sought.

The fact that these options weren't presented to me highlights a crucial issue in our education system: the need for comprehensive career guidance. Many students, like myself, miss out on opportunities simply because they're unaware they exist. It's a reminder of how important it is for educators and counselors to stay informed about all potential paths for their students, not just the most common ones.

Had I known about these academies, I would have pursued that route. It could have dramatically altered the course of my life and career. But without that knowledge, I decided to enlist in the United States Air Force, choosing the law enforcement career field for my training.

I also wanted to be a military working dog handler, but I quickly learned that this position wasn't available as an initial contract. After completing the law enforcement academy, I got my first taste of the often complex and sometimes frustrating nature of military career paths.

As I said earlier, my experience is far from unique. Many young people and well-meaning citizens face similar struggles when trying to serve their country or community, whether in the military or law enforcement. The entry process for these careers can be daunting, often involving rigorous physical tests, background checks, and, in many cases, educational requirements that catch applicants off guard.

The Armed Services Vocational Aptitude Battery (ASVAB) test can be a significant hurdle for military service. Many potential recruits scramble to brush up on academic skills they thought they'd left behind in high school. Similarly, police departments increasingly require college degrees or a minimum number of college credits, recognizing the complex nature of modern law enforcement.

While necessary for maintaining the quality and professionalism of our armed forces and law enforcement agencies, these requirements can discourage some applicants. I've seen many promising individuals give up on their dreams of service before getting started, intimidated by the academic requirements or complex application processes.

It's a stark reminder of the importance of education, not just for personal development, but as a key that unlocks doors to service and career advancement. Had I understood this earlier in high school, I would have approached my studies more seriously.

For those reading this who aspire to serve, whether in the military, law enforcement, or any other public service role, let my experience be a lesson. Embrace your education. Every subject, every class, and every book you read is building the foundation for your future success. You never know which knowledge might prove crucial in your career or serving your community.

And for educators, counselors, and mentors, I implore you to broaden the horizons of the young people you guide. Research and present all possible paths, even those seemingly less familiar. You never know which option might be the perfect fit for a student's aspirations.

My journey from an unmotivated high school student to a dedicated public servant was filled with unexpected turns and valuable lessons. While I may have taken a roundabout path, each step taught me something valuable. By sharing these experiences, I hope to help you navigate your journeys more effectively and make the most of every educational opportunity.

Remember, in military service and law enforcement, as in life, knowledge truly is power. Embrace it, seek it out, and let it guide you towards a life of meaningful service. You have limitless potential - all it takes is the decision to apply yourself and the willingness to see the bigger picture.

Chapter 4

Getting in the Hard Way

Despite the struggles I faced in my early education, my determination to serve in law enforcement never wavered. As I approached the end of high school, I realized I needed to find ways to prepare myself for this challenging career path. During this time, I discovered an organization that would be pivotal in shaping my future.

Looking back on the events surrounding my journey into law enforcement and military service, I'm reminded that the path to a life of service is rarely smooth. My struggles to break into my chosen field taught me valuable lessons about perseverance and adaptability—lessons that would serve me well throughout my career.

During my high school years, I joined the Civil Air Patrol (CAP), a decision that would prove pivotal in shaping my future. The Civil Air Patrol, established in 1941, is the official auxiliary of the United States Air Force. It's a unique organization that combines civilian volunteers with a military structure, focusing on three primary missions: aerospace education, cadet programs, and emergency services. It was the perfect training ground for a young man like myself, eager to serve and lead. I also loved the Air Force uniform; there was just something so cool about it.

The CAP cadet program is designed to foster leadership skills, physical fitness, and aerospace knowledge in young people aged 12 to 21. As I threw myself into the program, I found myself thriving in ways I never had in the traditional classroom setting. The structured environment, clear goals, and emphasis on personal growth resonated deeply with me.

I quickly rose through the ranks of the CAP cadet program. The CAP system combines aerospace education, leadership training, and physical fitness, requiring individuals to pass tests and demonstrate practical leadership skills for each rank promotion. I aimed to pass a monthly test and progress through the ranks as quickly as possible. Within my first year, I had worked my way through all eight enlisted ranks and started on the warrant officer ranks.

This rapid progression allowed me to earn the Billy Mitchell Award, a significant achievement in the CAP that would later allow me to bypass basic training when I joined the Air Force. But I didn't stop there. I continued to climb the leadership ladder, moving from flight leader to cadet squadron commander, then to group staff, and eventually to wing staff. I effectively oversee all cadet programs in Ohio and shoulder responsibilities that most adults would find daunting. It was a heady experience for a teenager.

The pinnacle of my CAP career came when I was elevated to wing commander—the highest cadet position in the state. It was an enormous responsibility for a teenager, as I coordinated 35 Squadrons spread across the state, but one I embraced wholeheartedly. My role was complex, as I balanced leading officers older than myself with respect for their ranks. I was promoted to cadet colonel, a rare achievement that filled me with pride and a deep sense of duty.

Working toward General Carl A was one of the most challenging and rewarding aspects of my time in the CAP was working toward General Carl A. Spaatz Award, often called the SPAATZ test. This is the highest award achievable by a CAP cadet and represents the pinnacle of our training program. The rigorous test covers leadership, aerospace education, and physical fitness. I took the test just before entering active duty in the Air Force. While there was later some confusion about the records, the experience of preparing for and taking that test was invaluable.

The Civil Air Patrol captivated me not merely with ranks and awards but with its compelling emergency services mission. CAP plays a crucial role in search and rescue operations, particularly locating downed aircraft. We were trained to use directional finding equipment to locate each plane's emergency locator transmitters. This hands-on, real-world application of our skills was thrilling and gave me a taste of what it meant to be of service in critical situations.

While our unit was never activated for a downed aircraft search during my time, we did serve in several local disasters. I vividly remember our response to a tornado that tore through our area and the massive blizzard that paralyzed northern Ohio for weeks in 1978. These experiences were eye-opening. I saw firsthand the devastating impact of natural disasters on communities and the incredible resilience of people coming together in times of crisis.

During the blizzard, our unit had two primary missions: We ran a shelter for displaced people without power or heat in their homes and provided crucial communications support to responders. This is where my interest in amateur radio came into play. Many of us in the CAP, myself included, were Licensed HAM operators. To become a HAM or Amateur Radio Operator, one must pass a comprehensive examination administered by the Federal Communications Commission (FCC) that tests the knowledge of radio theory, regulation, and operating procedures. My call sign was KA8EAK, a sequence of letters and numbers that became a part of my identity during those long hours of emergency communication.

I also held a radiotelephone license from the FCC, which allowed me to operate aircraft radios. This combination of skills - HAM and aircraft radio operations - proved invaluable during emergencies. We would shift between shelter operations and communications, establishing links with other HAM operators and emergency services organizations to coordinate logistics and manage medical emergencies.

During this time, I realized a profound truth that would guide the rest of my career: at a very young age, I found that being able to help other people was significant to me. The satisfaction of assisting those in need, of being a part of something larger than myself, was addictive. I knew then that whatever path my life took would involve service to others.

This realization led me to seek as much training as possible. When I was 16, the emergency medical technician (EMT) program was rolled out across Ohio. I was too young to take the class officially, but I could audit the 80-hour course and participate in all the practical exercises. We learned how to use various equipment types and provide specific aid to victims. It was challenging, but I was determined to absorb every knowledge I could.

Two years later, I completed the National Registry EMT course and successfully passed, earning my nationally recognized emergency medical technician certification. I also earned my advanced certification, which allowed me to administer IVs and insert esophageal airways. While I didn't pursue paramedic training—my heart was set on becoming a full-time police officer—I wanted enough medical training to help people effectively in emergencies.

Looking back, I see how all these experiences—the Civil Air Patrol, the disaster response work, the radio operations, the EMT training—prepared me for a life of service. These experiences taught me valuable skills, but more importantly, they shaped my character and worldview. I learned the importance of being prepared for anything. You never know what you'll face in emergencies, so having a broad knowledge base and skills is crucial. I also learned the value of clear communication, especially in high-stress situations. And I learned that true leadership isn't about rank or position—it's about taking responsibility and doing whatever needs to be done to help others.

As high school drew to a close, I faced a pivotal decision. I had met all the graduation requirements by the end of my junior year, and as I said earlier, the prospect of spending my senior year in study hall and early release seemed like a waste of time. I chose the latter when presented with the options of enrolling in college or enlisting in the military.

On August 22, 1979, at 17, I left home for my first trip on an airplane, bound for basic training at Lackland Air Force Base in San Antonio, Texas. My decision to join the Air Force was deliberate. I researched the different branches and their law enforcement roles. I had also carefully considered my options, knowing that I wanted a path that would best prepare me for a career in civilian law enforcement.

The Air Force's law enforcement specialty aligned most closely with civilian police work compared to other branches. Army military police primarily guard supply routes and operate prisoner-of-war camps, with limited traditional law enforcement duties. The Marine Corps was similar, focusing more on field duty than what they termed "white hat duty." In the Navy, you weren't eligible to be a master-at-arms in your second enlistment. However, The Air Force offered an experience similar to civilian law enforcement agencies, aligning perfectly with my career aspirations.

Arriving at Lackland Air Force Base, I faced an unexpected challenge. My achievements in the Civil Air Patrol allowed me to bypass much of my basic training, presenting an exciting yet pressure-filled opportunity. Before leaving for basic training, I received a copy of the Air Force Sergeants Association magazine. The headline on the cover read, "Hardest thing to do in the Air Force: bypass basic training." It wasn't a confidence booster, but I was determined to rise to the challenge.

My two weeks in basic training were unlike those of my peers. The first week involved in-processing, medical examinations, Inoculations and other administrative functions. In the second week, I had to complete every test other airmen took over their full basic training period, but I did it alone. It was a grueling and intense experience, but I could demonstrate I had the required knowledge to graduate from basic training.

As I completed these tasks, I couldn't help but reflect on how far I'd come from being an average student in high school. The discipline and study habits I'd developed in the Civil Air Patrol paid off in ways I couldn't have imagined. It was a powerful lesson in the value of perseverance and the importance of seizing growth opportunities, no matter where they come from.

After completing basic training, I moved to the Law Enforcement Academy, also at Lackland Air Force Base. The senior non-commissioned officer, noting my basic training bypass, decided to send me to the student leadership school. For two weeks, I underwent training that prepared me to be the class leader for the following law enforcement flight.

Leading my class through academic and scenario-based training challenges was a transforming experience. I found ways to serve my fellow trainees, ensuring everyone had the support they needed to excel. This commitment to collective success paid off – after our training, I was awarded the Honor Graduate Trophy. This recognition filled me with pride and reinforced my commitment to excellence in service.

My next goal was to become a canine handler, a specialized and highly competitive position. The rigorous selection process involved a physical fitness test to ensure candidates had the strength and agility to handle dogs and the right temperament to manage these highly trained animals. I was thrilled when I was selected to attend canine school.

The canine training program was an intense nine-week course in which I worked with a "green" dog—one with no prior experience or training.

Learning how these dogs were raised and trained to serve alongside military personnel was fascinating. The bond between handler and dog is unique and powerful, built on trust, consistent training, and mutual dependence.

My canine partner was Lobo J206, named for his ear tattoo number. Together, we worked tirelessly to meet the stringent certification standards for a patrol dog. The process was challenging but immensely rewarding. By the end of the training, Lobo and I had formed a strong partnership, and I felt a deep sense of accomplishment when we successfully certified her as a patrol dog.

With my training complete, I received orders to report to Ellsworth Air Force Base in South Dakota as a young man who had never been far from home, trading the possibility of an assignment in Myrtle Beach for the harsh climate of South Dakota was a shock.

I arrived late at night at the regional airport near the base, where my first experience was deplaning via a staircase rolled up to the aircraft rather than through a jetway. The small brick building serving as a terminal was sparsely furnished with no vendors. As a private pilot, I have used most fixed-based operator (FBO) facilities that offer more amenities. It was extremely different from what I had imagined. Following the instructions on a large metal sign for arriving airmen, I called the provided number and waited. An hour later, an airman dressed in a snorkel parka picked me up without a word of welcome, driving me to the visiting airmen's quarters in silence. It was an odd and somewhat unsettling introduction to my new home.

That first night was restless. Unable to sleep, I ventured into the day room and looked out the window. The sight greeted me was surreal—giant B-52 bomber tails illuminated by iodine lights and enormous jackrabbits, seemingly as large as my dog, moving across the snowy ground. It was the only TV channel broadcasting news about sow belly futures and other agricultural statistics. I began to wonder what I had gotten myself into.

Over the next few days, I was introduced to my squadron and assigned to barracks and a dog. The reality of my new role quickly set in as I was given a map and told about the eight-hour shifts. We would alternate between law enforcement and security posts every two hours during our challenging assignment of patrolling the weapons storage area in the depths of winter, trudging through snow two or three feet deep in temperatures well below zero. Walking the entire perimeter took about an hour and 45 minutes. On days when vehicles were scarce, I would be dropped off and left to stand in the open cold until my ride came to pick me up. It was far from the glamorous law enforcement career I had envisioned.

The equipment we worked with was also a stark reality check. My first police vehicle, assigned to me in February 1980, was a 1973 Dodge four-door pickup truck with a broken heater. The mechanics had removed the heater core, leaving a large hole in the firewall that we covered with cardboard and tape. It was a constant battle against the elements. The makeshift barrier would fail at speeds over 30 mph, letting in snow, rain, and frigid air. With no defroster and a panting dog fogging the windows, visibility was a constant challenge.

As a young airman, I had dreamed of lights, sirens, and conducting vehicle stops. The reality was far different. Our trucks lacked emergency equipment and didn't even have mounted radios. We carried old Motorola walkie-talkies with batteries that died within the first hour of our shift. I often found myself patrolling for hours without hearing any radio traffic, only to be flagged down and told I was being called on the radio.

My romanticized notions of a modern, well-equipped law enforcement agency quickly gave way to the reality of working with outdated, often unreliable equipment. However, this experience taught me valuable lessons about adaptability and resourcefulness.

Despite these challenges, my supervisors recognized my ambition, loyalty, and reliability - traits I honed in the Boy Scouts and Civil Air Patrol. They asked if I would be interested in becoming a bomb dog handler, a position that required being on call 24/7 to respond to any bomb threats on base. I accepted without hesitation, eager for the additional responsibility and challenge, and was assigned the base's only bomb dog.

A week before leaving for bomb dog training, I noticed my dog's eyes were glassy, and he seemed lethargic. My attention to detail - another skill sharpened by my earlier experiences - potentially saved my dog's life. The base veterinarian diagnosed a problem with the dog's spleen that required surgery. While this meant my dog couldn't accompany me to school, it earned me a letter of commendation for my observational skills.

Bomb dog handlers are certified as a team with their dogs, unlike narcotic dog handlers. This meant I needed a new partner. I flew back to Lackland Air Force Base to select a new "green" dog that had graduated from patrol dog school and been chosen for detector dog training. It took trying out eleven dogs before I found one that could pass the rigorous standards required for explosive detection work.

That dog was Lady, a "fast-burning" dog—intelligent, intense, and quick to learn. Over six weeks, I trained her in patrol tactics and explosive detection. By the end of our training, we were certified as both a patrol dog team and an explosive detection dog team.

Returning to Ellsworth Air Force Base, I resumed my law enforcement duties with the added responsibility of being the base's bomb dog handler. This new role came with additional constraints. Anytime I left my post, I had to call the law enforcement desk, informing them of my destination, how long I'd be there, and my expected return time. Remember, this was the early '80s – we didn't have cell phones or pagers. We relied entirely on landlines for communication.

As a teenager, I naturally wanted to enjoy nightlife with people my age. However, my responsibilities as a bomb dog handler severely limited these opportunities. Most of my time was spent on base, ready to respond to any potential threat.

During this time, the reality of military service truly hit home. Shortly after becoming a bomb dog handler, I had to report to the Central Base Processing Office to complete a will, life insurance forms, and other deployment-related documents. The gravity of potentially going to war – something I had never considered when joining the Air Force – suddenly became real.

As I filled out my will, I remember telling the clerk, "The only thing I own is a stereo and a car." I had only been at the base a few months and lived in the barracks – though, as the bomb dog handler, I was fortunate to have a single room due to my on-call status.

These administrative tasks brought back a childhood dream I'd had when I was about five to seven years old. In the dream, I was dying in a sandy environment, bleeding out. Now, facing the possibility of deployment to Iran, that dream took on a frightening new significance.

However, world events intervened. Operation Eagle Claw, a mission aimed at rescuing 52 American hostages held in Tehran, failed tragically when a helicopter crashed into a C-130 transport plane, killing eight U.S. service members. In the aftermath, we were taken off alert and returned to standard law enforcement duties. While relieved, I couldn't shake the feeling that I had come very close to a life-altering – or life-ending –

experience. I remember thinking, "Wow, that was very close," hoping it would be the extent of my brush with such high-stakes situations.

Yet, life in the military continued to provide moments of high tension and historical significance. On March 30, 1981, I was on duty when suddenly, rotating red lights at all intersections began flashing, and klaxon horns started blaring in buildings across the base. This was the signal for alert crews to scramble to their aircraft.

At first, I assumed it was just another routine drill. We often had exercises where crews would reach their aircraft and perhaps start them, but I'd never witnessed them taking off. This time was different. Every aircraft on the flight line soared into the sky with remarkable speed. Ellsworth Air Force Base, known as the "Showplace of SAC" (Strategic Air Command), housed B-52 bombers, KC-135 tankers for aerial refueling, and several missile sites. Seeing all these assets mobilize at once was awe-inspiring and somewhat terrifying.

About ten minutes later, I heard a radio broadcast announcing that President Reagan had been shot. I later learned that protocol dictated all air bases go to a high state of readiness if the president is injured, mobilizing aircraft to be ready for any potential missions.

Fortunately, John Hinckley's assassination attempt failed, and President Reagan recovered. But the incident drove home the real-world implications of our readiness training and the vital role we played in national security.

These experiences – the possibility of deployment, the constant readiness required as a bomb dog handler, and witnessing the total mobilization of SAC forces – shaped my understanding of serving. They taught me that service often requires sacrifice, whether giving up personal time, facing dangerous situations, or simply being ready to respond immediately. A life given to serving others may not always be comfortable or safe, but it's a necessary and noble path.

As I reached the end of my enlistment, I was 21 – the minimum age to be hired as a police officer. Eager to transition to civilian law enforcement, I returned home to Ohio. However, I quickly discovered that while 21 was the minimum hiring age, the average hiring age was 28. Departments had learned that especially for males, the maturity and right attitudes for law enforcement often didn't develop until the late twenties.

This setback was frustrating, but it didn't deter me. I continued to take written exams, physical fitness tests, polygraphs, and background investigations. I was determined to achieve my goal of becoming a police officer. It was a reminder that the path of service can be complicated and that persistence is often as important as ability. Therefore, in anything we do, we must be determined to succeed against all odds.

Chapter 5

The Fireman

Nothing about my path to becoming a full-time police officer was easy. The process in Ohio, like many states, relied heavily on the civil service system. This meant that every public safety job, whether in law enforcement or firefighting, was regulated by a board that screened applicants based on rigorous tests and evaluations.

The testing process was extensive and multifaceted. It began with a written examination covering general knowledge rather than specific law enforcement or firefighting information. You were invited back for a physical fitness evaluation if you scored high enough on this initial test. This typically involved navigating an obstacle course and completing a run, usually one or two miles in distance.

Completing the physical portion led to the next phases: a psychological evaluation, medical examination, and polygraph test. Once all these steps were completed, applicants were rank-ordered based on their overall scores, determining their hiring eligibility. I quickly became what felt like a professional test-taker during that period. Every Sunday, I'd eagerly grab the Cleveland Plain Dealer newspaper, scouring the help-wanted ads for police officer positions. No matter where the opportunity was in Ohio, I was ready to apply and begin the testing process.

My first experience with civil service exams came that summer when I sat for the Sandusky Fire Department and Police Department tests. These tests introduced me to Ohio's civil service system, which governs hiring for public safety jobs. I had never taken a civil service test before and wasn't entirely sure what to expect. I prepared as best I could using study guides in the public library.

The exams were held on consecutive days - Saturday for the fire department and Sunday for the police department. To my surprise, both tests were nearly identical; they focused on general knowledge rather than specific information about firefighting or law enforcement.

Several weeks after the tests, I received the results in the mail. To my astonishment, I had scored as the number one candidate for the fire department and placed sixth for the police department. The fire department had 18 openings, while the police department had none. This unexpected turn showed me that my path seemed clear, even if it wasn't exactly what I had envisioned.

The civil service rules required all agencies to maintain a current list of applicants, which meant testing was an ongoing process. I completed the additional testing requirements for the fire department while maintaining my top ranking throughout the process.

Interestingly, despite my top ranking, the fire department didn't immediately hire me. They knew my interest in law enforcement and seemed hesitant to bring me on board, perhaps fearing I might leave for a police job at the first opportunity. However, when they did offer me the position, I accepted, recognizing it as a valuable step in my public service career.

This decision led to a significant move for my family, as we relocated from the large urban City of Cleveland to the lakefront City of Sandusky, Ohio. Little did I know then how this move would shape the next two years of my life and provide me with invaluable experience in emergency services.

However, before serving in Sandusky, my path to becoming a firefighter began unexpectedly during a voluntary community service in the Air

Force, stationed at Ellsworth Air Force Base in South Dakota. Just beyond the gates of our base lay the small community of Box Elder, a place that would become instrumental in shaping my future as a fireman.

The Box Elder Fire Department was an all-volunteer organization, its ranks filled with active duty airmen, civilian contractors working on the base, and residents. My insatiable appetite for all things related to emergency management drew me to join their ranks. Looking back, I realize this decision was pivotal in broadening my skills and providing invaluable experience that would serve me well in my future role while serving as a fireman in Sandusky.

The training sessions were intense and eye-opening, revealing the complexities of firefighting that most civilians never consider. We learned crucial skills like how to handle a fire hose nozzle properly. It might seem simple, but slamming it shut too quickly can create a dangerous phenomenon called a water hammer - a sudden increase in pressure that can cause pipes to burst or injure firefighters. We were taught to close the nozzle gradually, allowing the water pressure to dissipate safely.

Another critical lesson was the importance of proper hose arrangement. A kinked or tangled hose isn't just an inconvenience; it can mean the difference between life and death in a critical situation. We spent hours practicing laying out hoses efficiently, ensuring a smooth water flow while keeping in mind that every second counts.

Emergency vehicle operations were another critical component of our training. Driving a fire truck is vastly different from operating a regular vehicle. The sheer size and weight of the truck and the urgency of emergencies require a unique set of skills. We practiced navigating tight turns, backing into narrow spaces, and making quick decisions under pressure.

The camaraderie among the volunteers was evident. Everyone was eager to help and learn from each other, creating an atmosphere of mutual support and shared purpose. I had found another family united in our desire to serve and protect our community.

Our calls often involved fires in mobile homes rented by military personnel. Many of these structures were older, with outdated electrical systems that could quickly spark dangerous fires. Each call was a stark reminder of the importance of our work and the lives at stake. The acrid smell of smoke, the intense heat, and the urgency of rescuing potential victims - these experiences were both challenging and deeply rewarding.

But firefighting wasn't my only passion. I had always been drawn to emergency medical services, fascinated by the ability to help people in their most vulnerable moments. Before joining the fire department, I had earned my National Registered EMT designation, and I was eager to use these skills in real-world situations.

One night, a call came in that would leave an indelible mark on my memory and teach me a lesson I carry to this day. An elderly pedestrian had been struck on the highway. I responded immediately, my heart racing as I drove directly to the scene in my vehicle.

When I arrived, one of our fire captains was already there providing care to the patient. He had removed his turnout coat - the heavy protective gear firefighters wear - and used it to cover the woman. The temperature was well below zero, the cold that seemed to freeze the air in your lungs, and his quick thinking seemed commendable at first glance.

I made a critical mistake that I hope others can learn from. Instead of conducting my thorough examination, I deferred to the captain's assessment. His rank and experience intimidated me, and I let that cloud my judgment. He outranked me and was much older, and I hesitated to question his actions. I monitored her vitals and waited for the ambulance to arrive, all the while leaving her covered with the coat and lying directly on the frozen ground.

When the ambulance team arrived, they quickly removed the coat, revealing the full extent of her injuries. The sight that greeted us was shocking. She had a compound fracture of her ankle, with bones protruding through the skin and a large pool of coagulated blood surrounding the wound. Moreover, she was likely hypothermic from lying directly on the frozen earth for an extended period.

The guilt I felt at that moment was overwhelming. I had failed to perform a basic head-to-toe examination, a fundamental procedure that could have dramatically improved her immediate care. This experience taught me a valuable lesson: regardless of rank or experience, when you're responsible for patient care, you must always trust your training and conduct a thorough assessment. The human life at stake is more important than any perceived hierarchy or authority.

I'm relieved to say that the elderly woman survived the accident and eventually returned home after an extended hospital stay. But the memory of that night stays with me, a constant reminder of the importance of following protocol and trusting one's instincts, regardless of perceived authority. Throughout my career, I've carried this lesson to teach others about the critical importance of thorough patient assessment in emergencies.

My time with the Box Elder Fire Department wasn't just about fighting fires and responding to medical emergencies. I also joined the Pennington County Emergency Rescue Squad, a diverse group of volunteers skilled in various specialized rescue techniques. From cave rescue to dive rescue, mountain rescue to vehicle extrication, this experience opened my eyes to the multifaceted nature of emergency response.

One particularly memorable training exercise involved cave rescue. As someone who had never been spelunking, the experience was thrilling and eye-opening. We entered a training cave, and once inside the main opening, we stripped down to T-shirts, blue jeans, and hiking boots. The constant temperature of about 55 degrees Fahrenheit surprised me—I hadn't realized that caves maintained such a stable environment.

We were equipped with helmets adorned with carbide lamps - the same type used by old coal miners. It was amazing how a flame the size of a cigarette lighter could illuminate an entire room underground in total darkness. The dancing shadows created an almost ethereal atmosphere, adding to the sense of adventure and slight trepidation.

As we ventured deeper into the earth, I encountered passages so narrow that even as a young, relatively thin man, I had to completely exhale to slither through on my chest between two massive pieces of rock. The thought that if the earth shifted just a slight bit, it could crush a person like a hot dog, with parts squirting out both ends, was terrifying and humbling. This experience gave me a profound respect for cave rescue specialists and their unique challenges. Trying to rescue someone in such confined spaces, especially if they were injured and immobilized, would be incredibly difficult and time-consuming.

Mountain rescue training was equally enlightening and challenged my preconceptions. When I thought of climbing a mountain, I had imagined a steep but manageable slope. Reality quickly dispelled this notion when we faced a 90-degree angle - a straight, sheer cliff rising about 150 feet into the air.

We were belayed for safety, but scaling this vertical face using nothing but our fingertips and toes to find purchase in the smallest cracks and crevices was daunting. The physical strain was intense; muscles I didn't know I had begun screaming in protest as we inched our way up the cliff face. Reaching the top was an exhilarating mixture of relief and triumph.

But the training didn't stop there. Once at the top, we immediately had to switch gears, retying our harnesses for rappelling back down. This rapid transition from climbing to descending emphasized the versatility required in real rescue situations, where conditions change instantly.

Water rescue training presented its own set of challenges, particularly for me. Growing up, my mother's fear of water prevented me and my brother from learning to swim correctly. To this day, I'm not a strong swimmer; I rely primarily on self-taught techniques. However, the rescue team's training proved invaluable, teaching me essential survival techniques.

We learned how to use pants or shirts as improvised flotation devices—a skill that could save lives in unexpected water emergencies. We also learned how to cup our hands and push down through the water to propel ourselves to the surface efficiently. Despite my limited swimming abilities, these practical, hands-on lessons gave me new confidence in water situations.

Throughout these diverse training experiences, my ultimate goal of becoming a full-time police officer never wavered. Each new skill I acquired, and each challenge I overcame was another building block in my journey toward law enforcement. These experiences in firefighting and specialized rescue operations prepared me for the multifaceted challenges ahead when the firefighter opportunity opened in Sandusky.

As I began my tenure with the Sandusky Fire Department, I entered the world of the "probie." This role came with challenges and responsibilities, many of which were far from glamorous but essential to the department's functioning.

Life as a probationary firefighter, or "probie," was a mix of excitement, hard work, and sometimes mundane tasks. Every Saturday was devoted to what we called "brass day." Our fire station, built in the late 19th century, was adorned with numerous brass fixtures that required regular polishing to maintain their shine.

Five brass poles were inside the station, stretching from the third floor to the apparatus room and from the second floor to the apparatus room. We would spend hours standing on ladders, meticulously polishing these poles until they gleamed. The handrails around the holes in the floor, as were the floor drains and doorknobs, were also brass. By the end of each Saturday, our arms would ache from the repetitive motion, but the sight of the gleaming brass throughout the station filled us with pride and accomplishment.

The daily routine of a Sandusky firefighter was structured and diverse. Our shifts began at 7 a.m. with a test of the alarm system - a klaxon sounded by the police department to ensure everything was in working order. This was followed by a briefing, usually conducted by the assistant chief or captain, covering upcoming inspections, hydrant tests, or specific maintenance tasks requiring attention.

We also engaged in ongoing in-service training on various pieces of equipment. This included working with ladders, the snorkel truck, and other specialized apparatus. These training sessions were crucial for maintaining our proficiency and ensuring we were always ready to respond effectively to emergencies.

I was assigned to the ambulance, which typically ran about six calls a day - usually four in the afternoon and two or three at night after we had gone to sleep. This kept me busy and allowed me to utilize my EMT skills regularly. However, the confinement of being unable to leave the station during shifts sometimes felt restrictive, especially compared to the freedom I had experienced in my previous roles.

After 5 p.m., we were allowed to relax, but we couldn't leave the station. We had lounge chairs on the first floor, a TV area with recliners on the second floor, and our assigned bunks. Despite these comforts, I couldn't shake the feeling of being somewhat imprisoned. I was there and working but couldn't come and go as I pleased.

The fire department consisted of three stations: Station 1 downtown, the main station, Station 3 near Cedar Point, and Station 5 on the far west end of the city. The location and nature of each emergency call would determine which station would respond, or in more severe cases, multiple stations would be dispatched together.

Despite feeling somewhat confined, I threw myself into the work. I attended the 200-hour Firefighter Academy in Columbus, Ohio, graduating as the honor student - an achievement that made our Chief, Chief Reed, particularly proud. This commitment to excellence would become a hallmark of my career, regardless of the uniform I wore.

As a Sandusky firefighter, I always kept sight of my ultimate goal of becoming a police officer. The firefighter's schedule - 24 hours on, 24 off, for three shifts, followed by four days off - allowed me to pursue part-time law enforcement opportunities while maintaining my firefighting duties. It was during this period that I worked in three separate police departments on a part-time basis, gaining valuable experience and adding to my resume.

Chapter 6

Journey into the Police Department

Before joining the firefighters, I had returned home with a burning desire to fulfill my lifelong dream of becoming a police officer. However, realizing this ambition proved more challenging than I had anticipated. Apart from not being up to the average hiring age, the economy in the early 1980s was in a slump, with steel mills shutting down and auto manufacturing taking a hit. This economic downturn had a ripple effect on local governments, making it difficult for cities to afford full-time police officers with benefits.

I applied to several police departments, eagerly awaiting a chance to serve my community. But rejection after rejection came each one a blow to my confidence but not to my resolve. I knew that somehow, I would achieve my lifelong goal, so I continued to take practice civil services exams to improve my scores and worked harder to improve my physical fitness to get my foot in the door of law enforcement.

Then, an opportunity presented itself that didn't come with a paycheck but offered something far more valuable: experience. The Twinsburg Township Police Department in Ohio was looking for reserve officers. These positions were unpaid, but they provided a chance to gain real-world law enforcement experience and the required training. For me, the lack of compensation was a non-issue. My passion for serving and protecting others far outweighed any financial considerations.

I threw myself into the application process, completing every form with meticulous care and acing the entry tests. My military background and unwavering determination must have shone through because on December 24th, 1982, while most people were hanging stockings and wrapping last-minute gifts, I was preparing for a different celebration. This wasn't just Christmas Eve; it was the day I would be sworn in as a part-time police officer.

This moment marked the beginning of my career in law enforcement, a path that would shape the rest of my life. As I stood there, taking the oath to serve and protect, I couldn't help but feel a sense of pride and excitement. Little did I know that this small township, about 27 miles from Cleveland, would provide me with experiences that would test my resolve, challenge my perceptions, and ultimately reinforce my commitment to service.

Twinsburg Township was primarily an African-American community with several subsidized housing projects within its jurisdiction. The police department was relatively new, having been formed under unique and harrowing circumstances just months before I joined.

To understand the context of my entry into the department, recount a few months before I was sworn in, the Summit County Sheriff's Office provided law enforcement services in Twinsburg Township under a contract. However, a terrifying incident prompted the sheriff to cancel this arrangement, necessitating the formation of the township's police force.

About six months earlier, a deputy had been dispatched to a disturbance call in the area. What should have been a routine response turned into a nightmarish scenario. Upon arrival, the deputy was overpowered by a group of men who managed to disarm him. In a moment of sheer terror, one of the assailants pointed the deputy's own .357 Magnum revolver directly at his head and pulled the trigger – not once, but six times.

In what can only be described as a miracle, the gun didn't fire. As sirens wailed in the distance, signaling the approach of backup, the assailants panicked, threw down the gun, and fled the scene. This incident, along with other assaults on deputies, deeply concerned the sheriff about the safety of his officers in this community.

The revolver that had failed to fire was sent to the Bureau of Criminal Investigation (BCI) in Ohio for testing. What they discovered defied explanation: the gun fired perfectly every time it was tested. The original six bullets that were in the gun during the assault all had dented primers and were fired on the first attempt in controlled conditions. There was no scientific explanation for why the gun had failed to fire during the attack on the deputy.

This inexplicable event was the sheriff's final straw. He gave the town a six-month ultimatum to form its police department. The community quickly assembled a committee and created a new force from scratch. They hired Dayton Brannon as their chief, along with a sergeant and two full-time patrol officers.

The new police station was a far cry from the high-tech facilities in larger cities. It was housed in the old fire department building, and the police cars were civilian vehicles painted to look like patrol cars. These weren't the specially equipped police package vehicles with reinforced suspensions, robust electrical systems, and powerful engines. They were regular cars dressed up to serve a noble purpose - much like how I, a novice officer, stepped into a role I had long dreamed of but was only beginning to understand.

It quickly became apparent that to cover 24-hour shifts, the department must rely on part-time and reserve officers. This is where my opportunity arose. I was one of the first reserve officers hired, which meant I was volunteering my time to work. While unpaid, I saw this as a golden opportunity to gain experience and the required training to launch my law enforcement career.

I owe a debt of gratitude to Officer David Wulff, one of the new full-time officers. He reviewed my application and, recognizing the value of my military experience, consistently placed it at the top of the pile for Chief Brannon's review. Dave's persistence played a crucial role in helping me start my law enforcement career, a kindness I've never forgotten that taught me the importance of champions and mentors in one's professional journey.

On the first night after being sworn in, I rode along with Dave. We responded to a domestic violence call, which, thankfully, ended without incident. It was a relatively quiet introduction to what would become a career filled with unpredictable and often dangerous situations.

Soon after, I began working on the housing projects within our jurisdiction. The Summit County Housing Authority had allocated funds to pay off-duty police officers to patrol these complexes. This arrangement was a win-win situation—it provided additional security for the residents and allowed the township to have two officers on duty simultaneously, enhancing officer safety through the ability to provide backup.

While my time at Twinsburg Township didn't involve any particularly memorable incidents, it gave me an invaluable opportunity – attendance at the Ohio Peace Officers Training Council Academy. This was the official police academy in Ohio, and the one I attended was run by the Cleveland Metroparks Rangers, a fully accredited law enforcement agency responsible for all the county parks in the Cleveland-Cuyahoga area.

Contrary to what some might assume, given the park ranger connection, this wasn't a course about identifying flora and fauna. It was a comprehensive police academy that covered everything from police procedures and physical fitness to law and practical exercises simulating real-world scenarios we might encounter on the street.

Graduation day from the academy remains one of my proudest moments. Standing in my crisp uniform, I couldn't help but beam with pride as I saw my mom and dad sitting in the audience. Little did I know then that this would be the only one of my many graduations they would attend, making the memory all the more precious.

As often happens, circumstances change, presenting new opportunities and challenges. During my time with Twinsburg Township, I got hired by the Sandusky Fire Department and moved from Cleveland to Sandusky. This relocation made the commute to Twinsburg Township quite challenging, leading me to seek opportunities closer to my new home.

I found another part-time position at Olmsted Township, Ohio, more conveniently near the highway. While I still occasionally worked shifts at Twinsburg, the paid position at Olmsted Township naturally became my priority.

Olmsted Township operated differently from Twinsburg. They maintained two officers on duty 24 hours a day and had a full-time police chief. However, they lacked a dedicated dispatcher. Officers alternated between dispatch duties and patrol work to address this, switching roles halfway through each shift. In emergencies requiring two officers, we'd temporarily enlist one of the full-time firefighters from next door to man the dispatch center.

It was at Olmsted Township that I encountered one of the most physically challenging situations of my early career – attempting to restrain a person under the influence of PCP (Phencyclidine). The incident began with a routine call – a single-car accident where the driver had run off the road into a deep ditch during winter.

Upon arrival at the scene, we found the driver injured but exhibiting bizarre behavior. Initially, we suspected alcohol impairment, but as we attempted to handcuff him, it became clear that we were dealing with something far more potent. The man seemed impervious to pain, displaying a strength and resistance that defied normal human limits.

After an intense struggle, we finally managed to secure him in the back of the patrol car with his hands cuffed behind his back. However, our relief was short-lived. In a display of drug-fueled strength and flexibility, the man managed to kick out the side door window of the police car. Then, in a move that still amazes me, he maneuvered his body so that he slithered out head-first onto the pavement.

We had to pick him up and put him back in the car, and I ended up having to physically sit on him to keep him restrained for the entire journey to the hospital. This experience was a stark reminder of the unpredictable and often dangerous nature of police work and the importance of continuous training to handle such extreme situations.

However, not all challenges in law enforcement come from confrontations. Sometimes, the threats are more insidious and long-lasting. An incident from my time back at Twinsburg Township is a chilling reminder.

Several outlaw motorcycle gangs operated in and around the Twinsburg Township area, including one known as the Iron Horsemen. I had observed them on club rides through our community on multiple occasions. One afternoon, I spotted a motorcycle belonging to one of their members with an expired registration plate.

Following protocol, I initiated a traffic stop. To my surprise, not just the individual I had targeted but the entire group of about eight to ten motorcycles also pulled over. I approached the motorcyclist I had initially spotted and asked for his driver's license. He informed me that his license was revoked. I arrested him and discovered the motorcycle's registration plate was fictitious.

I called for a rollback wrecker to tow the motorcycle, and once it arrived and loaded up the bike, I transported the biker to jail. The other bikers did not interfere, and the exchange remained professional. However, this encounter was far from over.

A few weeks later, as I left the police department, I noticed a motorcycle pull out behind me. I watched it as I approached the interstate to head home to Cleveland. The motorcycle followed me onto the interstate, maintaining a distance of about 10 to 15 car lengths behind me for the entire 40-mile journey into Cleveland.

As I exited the interstate near my neighborhood, the motorcycle followed. At this point, there was no doubt that this motorcyclist was tailing me, likely trying to discover where I lived. The implications were clear and deeply unsettling – this could be retaliation for the arrest I had made weeks earlier.

Knowing I couldn't lead this potential threat to my family, I made a split-second decision. Instead of heading home, I turned in the opposite direction, entering a different neighborhood. I found an old home with an empty driveway and an open garage. I pulled in, shut off my car, and positioned myself in the shadows, waiting to see if the motorcyclist would stop or turn into the driveway.

I watched as the motorcycle, which I could now identify as a Harley Davidson, rumbled past the street. After waiting several more minutes to ensure the coast was clear, I took a long, circuitous route home, constantly checking my mirrors for any sign of being followed again.

While I never saw that motorcycle again and never experienced any further negative encounters with the biker gang, to this day, I believe their purpose that night was to locate my home, potentially to do harm to me or my family when I was off duty. This incident served as a reminder of the potential dangers that can follow officers home long after their shift has ended.

As my experience grew, I continued seeking opportunities to expand my law enforcement career. This led me to another part-time position with the Plymouth, Ohio, police department. Like many small towns, Plymouth didn't have the financial resources to maintain a whole staff of full-time officers, relying instead on part-time employees to complete the schedule.

The new police chief in Plymouth, Steve Caudill, faced the challenge of reorganizing the police department. Chief Caudill was a good man who had come from a more prominent agency, and he was navigating the learning curve of managing a small-town department with limited resources and budget constraints.

While my time in Plymouth was relatively uneventful, one incident stood out, even though I wasn't directly involved. One of our officers responded to a call about a suspicious person and encountered a man wielding a machete. In a display of remarkable composure and de-escalation skills, the officer managed to talk the man down without drawing his weapon despite the man swinging the machete at him multiple times.

I remember Chief Caudill's reaction to this incident. He admitted, "I would have never been able to do that. I'm glad that he didn't have to shoot, but I don't think I would have been able to allow the guy to come at me with a machete, swinging it back and forth, without protecting myself." This incident highlighted the unpredictable nature of police work and the importance of de-escalation skills.

Reflecting on the early years of my law enforcement career, I am struck by the range of experiences and lessons I have learned from each employment. From the challenges of policing in a community with a history of tension with law enforcement to the resourcefulness required in small-town departments to the constant awareness of potential threats both on and off duty, these experiences played a crucial role in shaping my approach to policing and leadership.

However, as I continued striving to find my way into a full-time job in the police department, I quickly learned that the system, despite its intention to be fair, had its flaws. The most glaring example of this came during my experience with the Toledo Police Department in 1985. Toledo had 40 openings for police officers that year. After completing all the required tests, I was ranked number 11 out of all applicants. Given that position, I should have been a shoo-in for one of the available spots.

Civil service rules typically state that the employer must select one of the top three candidates on the list at any time. This meant that even if you were number three, you might be chosen first and hired. Then, candidates one and two would move down the list. They would be removed entirely from the list if passed over three times.

So, you can imagine my shock when, during my final criminal background check, the Toledo detective conducting the interview bluntly told me, "You know you're not getting this job, right?" When I expressed my confusion, given my high ranking, he explained that due to diversity initiatives, they had to consider every African American, female, Hispanic, and other minority candidate before even looking at white male applicants.

This was my first encounter with what I perceived as reverse discrimination, and it was a harsh reality check. The detective informed me that the only portion of their test not contestable in court was the psychological exam. Sure enough, a few days later, I received a letter stating that I had not met their requirements for a psychological profile.

The irony of this situation wasn't lost on me. I underwent a similar psychological evaluation during my firefighter training with the Sandusky Fire Department. In that instance, the psychiatrist had told me that my profile was typical of a police applicant - showing good judgment, honesty, and integrity. The only caveat was a note that I might have a propensity for substance abuse, though I neither smoked nor drank regularly.

This experience with Toledo left a bitter taste in my mouth. When they called me back months later, asking me to retake the psychological portion of the test, I declined multiple times. In hindsight, I sometimes regret this decision. Toledo was a large department with all the specialized units that a young officer might dream of. You name it: helicopters, boats, canine units, mounted police, Toledo had them all. However, my disappointment in how I was initially treated outweighed the potential opportunities.

These experiences taught me valuable lessons about the sometimes complicated nature of pursuing a career in law enforcement. They also reinforced my belief in the importance of fair and equitable hiring practices. Throughout this time, I always kept sight of my goal to become a full-time police officer. Despite these setbacks, I continued to take tests and seek opportunities, knowing that law enforcement was my true calling. While I enjoyed the part-time jobs, my work as a firefighter, and the satisfaction of helping people in medical emergencies and accidents, I knew that my skills and passion were best suited to law enforcement.

In the spring of 1986, my persistence paid off when I was offered a position with the Cary Police Department in North Carolina. I completed all their testing requirements and was given a start date of July 6th, 1986. With my wife and two children, I got ready to move south and start another chapter in my life of service.

However, this meant I had to resign from my other job as a firefighter. My last day as a Sandusky firefighter perfectly encapsulated the diverse and challenging nature of the job. First, it began shortly after our morning roll call with a severe car accident. A driver had lost control and collided head-on with an iron beam supporting a railroad trestle. The impact was so severe that the unbelted driver was entirely on the floorboard, his head resting on the seat.

This complex extrication required all of our skills. We started IVs to stabilize the patient while calling for our rescue truck. We had to cut off the roof, peel it back, pull the steering wheel up and out of the car, and use spreaders to open the door panel. It was a painstaking process, but we finally packaged the patient and loaded him into the ambulance.

Later that same day, we responded to another motor vehicle crash. An elderly lady had run a stop sign and collided with a motorcyclist, severely injuring him. The situation called for immediate action. I used 10-gauge needles - large enough to see through - to start IVs in both arms, running fluids at full speed to replace the blood he was rapidly losing. It was a race against time, but we managed to stabilize him, get him packaged up, and transport him to the hospital, ultimately saving his life.

As night fell on my final shift, I went to bed, hoping for a peaceful night's sleep. However, the fire alarm had other plans. We were roused from our bunks for a fully involved house fire. Although I responded with the ambulance, I was reassigned to the pumper and told to assist with roof ventilation.

Ventilating a roof is a crucial but dangerous task in firefighting. It involves cutting through the roof to release heated gases and smoke, allowing firefighters inside to work more effectively. Armed with a K-12 saw—essentially a chainsaw with a circular blade—we ascended to the roof of the three-story home.

I climbed the ladder and noticed it wasn't set at the correct angle. We were taught that if you stand on the bottom rung and place your hands directly in front of you, you should be able to extend your arms fully.

This ladder was set too steeply. Despite my concerns, I followed orders and began climbing with a heavy inch-and-a-half hose attack line.

Halfway up, disaster nearly struck. The ladder began to tip backward, threatening to send me plummeting three stories to the ground. In that heart-stopping moment, all I could think about was the weight of my gear - the turnout coat, the self-contained breathing apparatus, the face mask, the helmet - and how it would likely snap my neck if I fell.

Fortunately, my fellow firefighter on the roof saw what was happening. In a split-second decision that likely saved my life, he grabbed the nozzle of the hose I was carrying and pulled the ladder back towards the house, allowing me to scramble to safety. It was a stark reminder of the daily dangers and the importance of looking out for one another.

We ventilated the roof, extinguished the fire, and spent the remainder of the shift cleaning hoses and repacking the trucks, ensuring everything was ready for the next call. It was a fitting end to my time as a firefighter—a shift encompassing all aspects of the job, from lifesaving medical interventions to battling a raging inferno.

As I hung up my turnout gear for the last time, I reflected on the wealth of experiences I had gained. The skills I had honed - from emergency medical care to technical rescue operations - would prove invaluable in my career as a police officer. More importantly, I left with a deeper understanding of the importance of teamwork, quick thinking, and unwavering commitment to public safety.

My journey as a firefighter might have been relatively brief, but its impact on my life and career was profound. It reinforced my passion for public service and prepared me for the challenges in law enforcement. As I closed this chapter of my life, I knew that while leaving behind one form of service, I was moving towards another equally vital role in protecting and serving my community.

Chapter 7

From Part-Time Police Officer to Police Chief (Part 1)

It was July 1986 when I first pinned on that badge as the 37th police officer in Cary, North Carolina. Cary was no ordinary town; it was on the brink of transformation, nestled in the heart of the booming Research Triangle Park. The air buzzed with the energy of tech startups, medical breakthroughs, and cutting-edge research. As a young officer, I could feel the pulse of progress beating beneath my feet.

Cary was a town of about 45,000 people back then, a far cry from the bustling city it would become. The police department, with its 37 full-time officers, seemed substantial at the time. However, their grand vision for the future captured my imagination during the interview.

"By the year 2000," they told me, their eyes alight with enthusiasm, "Cary will be home to over 100,000 people, and our police force will grow to more than 100 officers."

I leaned forward in my chair, captivated by the potential. They spoke of an inner beltway named Maynard Road and an outer beltway called Cary Parkway. To a young officer hungry for opportunity, this was more than just a job offer—it was a golden ticket to a future filled with possibilities.

The promise of growth meant specialty positions, supervisory roles, and a chance to climb the ranks. It was a future I could see myself in, a career path stretching out like an open road. Little did I know then how prophetic those projections would prove to be. By 1999, all those predictions had come true, transforming Cary and its police department in ways I could scarcely have imagined on that summer day in 1986.

The department exuded an air of professionalism that resonated with me. Most officers welcomed me warmly despite my status as a newcomer from up North. There was a sense of camaraderie, a shared purpose that transcended regional differences. It felt like a place where I could belong and make a difference.

However, my journey with the Cary Police Department wouldn't begin on the streets. As I wasn't yet a certified police officer in North Carolina, my first stop would be the basic law enforcement training at the North Carolina Justice Academy in Salemburg. But before that, the department had a unique approach to introducing new officers to the force.

Along with two other recruits, I was assigned to rotate through different divisions of the police department. Each week, we'd immerse ourselves in a different aspect of law enforcement operations. We spent time in investigations, learning the intricacies of detective work. We sat alongside dispatchers, understanding their critical role in coordinating police response. We even shadowed the dog catcher, gaining insight into animal control operations.

This comprehensive exposure was more than just an orientation—it was a valuable education in the interconnected nature of police work. By the time we were ready to hit the streets, we'd have a holistic understanding of how each cog in the machine contributed to the department's overall function.

As the day approached for us to head to the North Carolina Justice Academy, rumors began circulating. Word around the department was that one of the new officers—a North Carolina State graduate who had interned with us—was a shoo-in for success. The other two rookies? Well, the jury was still out on us.

The drive to Salemburg was filled with anticipation and nervousness. As we pulled up to the academy, we were greeted by our school coordinator, Earl Hardy. Earl, with his imposing presence and no-nonsense demeanor, had spent most of his career with the North Carolina Highway Patrol. He had left that position as the physical fitness instructor for their academy, which told us one thing—we were in for a physically demanding experience.

When we stepped onto the academy grounds, it felt like we had been transported into a military boot camp. We stood at attention in crisp formations, our uniforms meticulously pressed and our boots shining. Inspections were rigorous, with every detail scrutinized. We marched everywhere, our footsteps echoing in unison across the campus. And we ran—oh, how we ran. Every movement seemed to be at double time, pushing our bodies to their limits.

Earl Hardy's words hung heavy in the air during the orientation class. "Look to your left," he commanded, his voice carrying the weight of years of experience. We obeyed, eyeing our fellow recruits. "Now look to your right." Again, we complied. "Now look forward," he continued, his gaze sweeping across the room. "Half the people in this room will not graduate."

The gravity of his words settled over us like a thick fog. This wasn't just a training program—it was a crucible designed to test us to our core. Those lacking mental and physical fortitude would be weeded out, leaving only those genuinely fit to wear the badge.

On just the second day of the academy, a whisper reached my ears that would shake my confidence in its foundation. The officer who had been pegged as a sure thing—the one with the internship and the prestigious degree—leaned over to me and muttered, "I'm leaving."

At first, I didn't grasp the whole meaning of his words. In my naivety, I thought he might be planning to sneak off campus for a quick lunch or to grab some contraband snacks. After all, we were essentially confined to the academy grounds, forbidden from venturing into the small community of Salemburg for any reason.

But as the reality of his statement sank in, I realized the enormity of what was happening. This wasn't about sneaking out for a moment's respite—he was resigning. The pressure of the academy, with its military-style discipline and rigorous expectations, had proven too much for him.

The inspections were particularly grueling. Staff members would storm into our quarters, their eyes scanning for imperfection. If a bed weren't made to exacting standards, the mattress would be unceremoniously tossed out into the yard. Personal hygiene items, socks, underwear, and t-shirts had to be displayed precisely in our drawers. Any deviation from the expected layout would result in our belongings being strewn across the floor.

For some, like our departing colleague, this scrutiny and constant pressure was too much to bear. It reminded me of the mental fortitude required to complete the academy and thrive in the high-stress world of law enforcement.

As I watched him leave, a mix of emotions washed over me. I felt sympathy for his struggle and a steely determination that took root in my gut. If he couldn't make it, we had to work even harder. We had to prove that we were made of tougher stuff.

With renewed focus, I threw myself into every aspect of the training. The physical demands were grueling, but I pushed myself beyond what I thought were my limits. The academic portions were challenging, but I studied with an intensity I'd never known. Every inspection became a personal challenge to exceed expectations.

As the weeks passed, I found myself surviving and thriving in this high-pressure environment. The discipline, attention to detail, and constant push for excellence shaped me into the officer I aspired to be.

When the final tests came, I was ready. I poured everything I had learned, every ounce of determination I possessed, into those exams. And when the results were announced, I stood tall, proud to be named the honor graduate of the academy.

The other officer from Cary, Steve Modlin, who had stuck it out with me, also performed admirably, finishing not far behind in the class standings. We had weathered the storm together, emerging stronger and more capable on the other side.

As we prepared to return to Cary, I couldn't help but reflect on our transformation. We were no longer wide-eyed recruits but trained professionals ready to take on the responsibilities of protecting and serving our community.

In November 1986, we stood before our fellow officers in Cary, raised our right hands, and took the oath to become sworn police officers. The weight of the badge, pinned to my chest, served as a constant reminder of the challenges we had overcome and the duty we now bore. Little did I know then that this was just another junction in my journey in law enforcement and that I was about to discover that the real tests of character and leadership were still ahead, waiting for me on the streets of Cary and beyond.

Fresh out of the academy, I was eager to hit the streets of Cary and put my training into practice. However, my first experiences as a sworn officer were far from what I had envisioned during those weeks at the North Carolina Justice Academy.

I was assigned to a patrol squad, but my introduction to real police work was delayed. Our Sergeant was on leave, deer hunting in some distant woods, leaving our squad leader in charge and acting as my field training officer. I had high hopes for comprehensive, hands-on training, but I got a crash course on the less glamorous side of police work.

My first night shift started with promise. We left the police department, the air thick with anticipation – or so I thought. Our first stop? The squad leader's home, where we settled in for a pot of coffee. The bitter aroma filled the air as I waited, wondering when the real work would begin.

Next, we headed downtown to a local restaurant. As we nursed more coffee, I realized we were there for a specific purpose – waiting for the squad leader's girlfriend to close the shop. We followed her to the bank for the nightly deposit, and my training consisted of being an unofficial security detail.

The rest of the night was a blur of caffeine and aimless patrols. We circled back to the squad leader's home for yet another pot of coffee, then cruised around town before ending up at the 24-hour Waffle House. As I sat there, jittery from the endless cups of coffee and lack of action, I couldn't help but wonder if this was what being a police officer was all about.

When my shift ended at 7 AM, I was wide awake, buzzing with unused energy and unfulfilled expectations. The next few nights followed a similar pattern, leaving me questioning the effectiveness of my "training."

Day shifts, however, were a different story. Cary's rapid growth had outpaced its infrastructure, leading to heavy traffic and frequent accidents. On average, I wrote four to six traffic crash reports daily. The calls for service picked up, too—larceny cases, property damage reports, and other incidents kept us busy throughout the day.

One particular day stands out in my memory. The traffic lights at a main intersection malfunctioned, and my squad leader and I took it upon ourselves to direct traffic. For nearly our 12-hour shift, we stood in the middle of that intersection, arms waving, sweat pouring down our backs in the November heat. It was exhausting, monotonous work, but it gave me my first real taste of what it meant to serve the community.

As we transitioned back to night shifts, I was paired with a different officer – one known for his proactive approach to policing. This, I thought, was my chance to learn the ropes. We were working the DuPont schedule: four-day shifts, two days off, three-night shifts, two days off, three-day shifts, two days off, four-night shifts, and a glorious seven days off.

On one of these night shifts, with me behind the wheel, we spotted a car weaving erratically on the road. The telltale signs of a drunk driver were unmistakable. Heart pounding, I activated the emergency lights and conducted my first traffic stop.

The smell of alcohol hit me as soon as the driver's window rolled down. Following the procedure, I asked the driver to step out of the vehicle and administered the field sobriety tests: the walk-and-turn, the one-leg stand, and the horizontal gaze nystagmus. As the driver stumbled through each task, I felt a mix of adrenaline and certainty. This was it, my first drunk driving arrest.

I read his rights and secured him in the back of our patrol car. Back at the station, my partner took the time to critique my performance. He pointed out areas where I could improve my officer safety tactics and gave me tips on being a more effective officer. It was the kind of practical, hands-on training I had been craving.

The next night, I was released from field training and sent out alone. I felt excitement and trepidation as I patrolled the streets of Cary alone for the first time. Little did I know, I still had so much to learn.

As the months passed, I quickly realized that no two days would be the same. Each call presented a new challenge and opportunity to make a difference with my service in Cary, from the mundane to the extraordinary.

From day one, the job was a whirlwind. One minute, I might coax a frightened possum off someone's porch, much to the relief of panicked homeowners. The next, I could be in hot pursuit of bank robbers, adrenaline coursing through my veins as we raced to bring justice to our community. This diversity of experience was both exhilarating and humbling, a constant reminder of the multifaceted role we played in the lives of Cary's citizens.

But it wasn't always a high-stakes drama. Much of police work, I learned, was about persistence, patience, and attention to detail. This lesson was driven home in the summer of 1988 when our town faced a peculiar crime wave that tested our resolve and resourcefulness.

It started innocuously enough – reports of stolen cars began trickling in, nothing unusual for a growing suburb. But soon, a pattern emerged that set this spree apart. The thieves weren't just stealing cars; they were embarking on rampages of wanton destruction that left our peaceful neighborhoods in disarray.

Imagine a quiet residential street lined with lawns and mailboxes at the end of each driveway. Now imagine a stolen car, driven by reckless hands, jumping the curb and careening down the street, knocking over every mailbox and planter in its path. The destruction was methodical, almost ritualistic in its thoroughness.

For the victims, it violated their sense of security, a harsh reminder that chaos could strike even in our idyllic community. For us officers, it presented a logistical nightmare. A single night of vandalism could eat up an entire shift, going door-to-door, documenting damage, and trying to reassure shaken residents.

Our report management system, cutting-edge for its time, still relied heavily on physical paperwork. We carried massive three-ring binders filled with form templates, each designed to capture specific details for our database. What might seem like a simple act of vandalism to an outsider translated into hours of meticulous documentation for us?

As the weeks wore on and the incidents multiplied, the toll on our department and the community grew. The criminals grew bolder, escalating from property damage to more dangerous acts. In one harrowing incident, they even aimed a stolen car at the front of a house, causing severe structural damage as it breached the wall. Another time, they attempted to set a car ablaze after siphoning its gas, a chilling escalation that thankfully failed.

The pressure was mounting, and we all felt it. That's when Chief David Fortson, determined to end this crime spree, made an announcement that sparked a fire in all of us: he would treat the officer who caught these suspects to a steak dinner. It wasn't about the meal but a challenge and a call to action. It was about proving to ourselves and our community that we could overcome this challenge.

I took that challenge to heart, redoubling my efforts to catch these elusive criminals. I had no idea that my opportunity would come on what started as a routine morning patrol.

That morning, I was en route to a routine call when something caught my eye – a small Toyota pickup truck weaving erratically on the wrong side of the road. My instincts kicked in immediately. Was it a drunk driver? A medical emergency? Whatever the case, I knew I had to investigate.

I flipped on my blue lights and siren, executing a quick U-turn to pursue the vehicle. As I drew closer, I saw two white males in the cab. But instead of pulling over, the truck accelerated, and suddenly, I found myself in a high-speed chase through the streets of Cary.

My heart was pounding as I radioed in my situation and location while keeping my eyes locked on the fleeing vehicle. Other units began converging to assist. When it seemed they might slip away, the truck made a fatal error as it turned down a dead-end street, leaving the suspects nowhere to go. As the pickup skidded to a halt, both doors flew open, and the occupants bolted into the surrounding neighborhood.

"Dispatch, suspects fleeing on foot," I called in. I was out of my cruiser in a flash, giving chase through a maze of backyards. My heart pounded as I vaulted fences and dodged clotheslines, always a step behind. But determination and training paid off as I rounded a corner. I spotted one of the suspects trying to squeeze under a wooden patio deck barely inches off the ground.

With my weapon drawn, I ordered, "Come out slowly. " " Hands first, and they better be empty."

Slowly and cautiously, the suspect emerged. To my surprise, I found myself face to face with a teenage boy, not a hardened criminal but a teenage boy. He wore jeans and a T-shirt, supported by an incongruous women's web belt holding a large hunting knife. His leather gardening gloves completed his odd ensemble—a naive attempt to avoid leaving fingerprints.

As I secured him in handcuffs, my mind raced at the thought that he was just a kid, barely old enough to drive. What drives a kid to this kind of destructive behavior? But there was no time for philosophy - we had another suspect to catch. Where was his partner? The answers would come, but not before another shocking revelation.

Back at the station, as I began to unravel the identities of our suspects, I made a discovery that hit close to home for many in law enforcement. The second suspect, still at large, was the son of a supervisory special agent with the North Carolina State Bureau of Investigation. Making that phone call - waking up a fellow law enforcement officer to tell him his son was implicated in a month-long crime spree - was one of the hardest things I'd ever had to do.

But it drove home a crucial lesson: the law applies equally to everyone, regardless of your parents or societal position.

I called the agent at home and woke him up. I asked him if he was the suspect's father. He said he was. I asked him if he knew where he was, and he said he was asleep in his room. I said, "Would you please go check." He complied, came back, and said he was not there. I said, "Well, when he gets home, would you please give me a call and bring him to the Cary Police Department?" I then told him the range of charges that his son was facing.

To his credit, the agent brought his son in later that day. The boy was given his juvenile Miranda warning, and his father was present. They decided not to answer questions and left without any admission of guilt. I filed juvenile petitions against the two young men for the hundreds of mailboxes they had knocked down, for the home they had destroyed, and for the number of cars they had stolen and damaged.

It was a lengthy list of criminal activity over a short time. Eventually, the cases were consolidated, and the boys were found to be delinquent. Their parents had to pay restitution for all the damages their children had caused.

Chief Fortson offered to take me out for that steak dinner as promised. In a move that surprised him, I suggested breakfast at a local spot instead. It was a valuable opportunity to connect one-on-one with our new chief, a reminder that in law enforcement, building relationships is just as meaningful as enforcing the law.

This case taught me the importance of integrity in service. As police officers, we must be unbiased, holding everyone to the same standard regardless of their background or connections. It's not always easy, but it's essential for maintaining public trust and the justice system's integrity. However, this case also highlighted another vital lesson about service: balancing our professional lives with our responsibilities.

For that supervisory special agent, waking up to learn that his son had been implicated in a crime spree must have been devastating. It's a sobering reminder that while we dedicate our time and energy to serving our communities, we must not neglect the lives of those closest to us. The pressures of law enforcement—or any demanding job—can pull us away from family, making it easy to overlook the struggles or needs brewing at home.

As people in service roles, we're often called to go above and beyond for others, but in doing so, we sometimes leave our own families behind, thinking they'll be okay without our attention. No matter how committed we are to our work, our relationships outside need just as much care and attention. That agent's son, wrapped up in destructive behavior, may have sent signals long before the crime spree started that went unnoticed in the bustle of his father's duties.

This isn't to blame the agent; we've all been there. Service demands can blur the line between professional and personal life, making it hard to see the warning signs in our homes. But it's a reminder that while we serve others, we must also serve our families, making time for connection and ensuring they feel supported and seen.

Maintaining that balance is difficult, but it's a necessary part of living a life of service. If we neglect our personal lives, the consequences can ripple out, affecting our families and our ability to serve the community with a clear conscience and open heart. We must remember that our service doesn't end when we clock out for the day—it extends into our homes and relationships, and we must give them the same commitment we give to our roles in public service. Only then can we be whole, effective servants of our communities and families.

Chapter 8

From Part-Time Police Officer to Police Chief (Part 2)

Moving from a part-time officer to a full-time role at the Cary Police Department was a significant step forward in my career, but it came with its trials. The field training was short, maybe too short—I was only required to complete two-day shifts and three-night shifts before I was cleared to patrol on my own. My field training officer was confident, telling the supervisor, "Roy's like a pilot. He knows what to do; he just needed a check-out in the new aircraft, and now he's good to go."

Though I appreciated the vote of confidence, looking back, I wish training had been more thorough. Cary was larger and far more active than the small Ohio towns I'd been used to, and the level of criminal activity was on a whole different scale. A more extended training period would've given me more time to adjust to the increased demands and, more importantly, the difference in policing styles between the North and South.

I'd grown up in what's often called the "Dragnet era," influenced by the cold, efficient style of policing made famous by shows like *Dragnet*. In the North, police officers were direct, maybe even a little rough around the edges. We focused on the facts, handled business quickly, and moved on to the next call. That worked up north, but in the South, things were different. Policing was more personal and more polite. Citizens expected officers to do their jobs and engage them with courtesy and respect, which wasn't ingrained in my approach.

Roy Taylor Meeting with Citizen

At first, I struggled. My straightforward, no-nonsense style rubbed people the wrong way, and I started receiving complaints—not for misconduct but for what people saw as rudeness. "I need to see your license and registration," I'd say, thinking nothing of it. But in Cary, that kind of abruptness came off as cold or impolite, and it didn't take long for me to realize that if I wanted to succeed in this community, I needed to adapt.

One of the most important lessons I learned in those early days was the value of showing respect—both to the people we serve and to the people we serve with. Service isn't just about upholding the law or maintaining order. It's about treating people with dignity, making them feel heard, and understanding that how you deliver your message is as important as the message itself. When I started slowing down, offering people a chance to express themselves, and showing humility, situations that could have escalated began to de-escalate. I was still getting the job done, but now I was doing it in a way that built trust and respect.

As I found my Carry rhythm, I noticed an odd trend. Every time a new supervisor was promoted, they'd be assigned to my squad. It was a strange coincidence, but it wasn't always welcome. These new supervisors, many of whom were promoted based on seniority rather than leadership ability, seemed to see their role as "policing the police." They implemented new rules and nitpicked minor details, making the atmosphere uncomfortable.

I had heard whispers from veteran officers that the real stress in law enforcement didn't come from dealing with criminals but from navigating the department's administration. As I was under the microscope of these new supervisors, I began to understand what they meant. One supervisor instituted a policy requiring us to radio for permission before taking a meal break. We also needed separate permission to leave our patrol zone to eat. Given the limited dining options in Cary at the time, this often meant jumping through hoops to grab a quick bite.

The supervisor would often stake out the restaurants, timing our arrivals to ensure we weren't sneaking in early meals. As our 30-minute break neared its end, we'd see the supervisor's black and white unit circling like a shark, ensuring we were back in our patrol cars on time. These overbearing policies led to officers making mistakes out of fear of reprimand. One night, my friend and fellow officer, Robbie McGowan, made several errors that could have had severe consequences.

Robbie, always dreaming of becoming a North Carolina State Highway Patrol Trooper, was meticulous about his appearance. One night, he stopped in a parking lot to adjust his uniform. In his quest for perfection, he made a critical error – he removed his gun from its holster and placed it on the trunk of his patrol car while he tucked in his shirt.

In his enthusiasm to emulate the State Highway Patrol, he had also adopted their habit of turning the siren in his patrol car to the radio selector knob. This would rebroadcast any radio traffic over the PA speaker, a practice the Highway Patrol used because they didn't have portable radios. For Robbie, it was just another way to feel more like the trooper he aspired to be.

As fate would have it, a call came in for a loud music complaint. Robbie, knowing that answering on his portable radio might raise questions from our overzealous supervisor about why he was out of his patrol car, quickly jumped into the driver's seat. He acknowledged the call on the mobile radio and sped off to the apartment complex where the complaint had been lodged – forgetting his gun on the trunk.

It wasn't until he arrived at the scene and reached for his weapon that he realized his mistake. In a panic, he quickly dealt with the noise complaint and raced back to the parking lot, praying his gun would still be there. To his horror, it was gone.

He then tried to radio the sergeant, but because he had set his radio to broadcast, it had somehow gotten stuck between that and the manual button, preventing him from transmitting. Desperate, he walked to a nearby pay phone and called 9-1-1, reaching our dispatcher, David Clinger.

"I've lost my gun," Robbie confessed, his voice shaking.

Understanding the gravity of the situation, David responded calmly, "Don't call the sergeant. I'll get Roy and some others to come up there. We'll find your gun."

Soon, a small group of us were scouring the area, our flashlights cutting through the darkness as we searched for Robbie's missing weapon. After what felt like an eternity, we found it. The gun had slid off the trunk on Maynard Road, hit the curb, and suffered some damage. The yoke was bent, preventing the cylinder from opening, and the crane was twisted, rendering the weapon inoperable.

Robbie worked the rest of his shift with a gun that couldn't fire and a secret weight on his shoulders. We contacted our friend Officer Dave Wulff, one of Cary PD's armorers, who repaired the gun with spare parts without the supervisor ever finding out. This served as a reminder of how quickly things could go wrong, even with the best intentions.

The experience also highlighted the unintended consequences of over-regulation and the stress it could create. Robbie's fear of being disciplined for being out of his car led to many decisions that could have had more severe repercussions. It was a lesson in how seemingly small rules could sometimes lead to more significant problems.

Not long after, I had my brush with this overzealous supervision. One night, while checking the perimeter of the Kmart store on Kildare Farm Road, I spotted a man straddling the 12-foot-high fence around the garden center, clearly about to drop in. I challenged him, and he jumped off the fence and started running.

I was in my 1988 Chevrolet Impala police car, windows down as usual so I could hear better. As I started to give chase, I had to multitask - raising the windows, radioing for assistance, switching off the ignition, grabbing the key, shutting and locking the door. In my haste, the windows hadn't fully closed. There was a large gap for a determined person to reach in and possibly unlock the door.

But there was no time to dwell on that now. I took off after the suspect, chasing him into the woods. I immediately lost sight of him. in the dense foliage, but I pressed on, straining my ears for any sound that might give away his location. The woods were eerily quiet, and after several minutes of fruitless searching, I reluctantly returned to my patrol car.

To my surprise, I found the squad leader sitting beside my vehicle. He hadn't gotten out to assist me in the woods or even called out to let me know he was there. Instead, he asked, "Did he lay down on you?"

"I think so," I replied, still catching my breath. "I can't find him."

His response caught me off guard. "Okay, well, come on up to the office."

Confused, I asked, "What do you mean come up to the office?"

"Well, I'm going to write you up," he said matter-of-factly. "You left your car unsecured."

I was dumbfounded. Here I was, having just attempted to apprehend a suspect in an active crime, and my supervisor's primary concern was that I hadn't waited those extra few seconds for my windows to roll up completely. It seemed absurd, but I knew arguing would only make things worse.

As I sat in his office, receiving my official reprimand, I couldn't help but think about the message this sent. Were we more concerned with following every minute procedural detail than catching criminals? I understood the importance of securing our vehicles - we carried shotguns in gun sleeves on the floor in front of the driver's seat, after all. But surely, there had to be room for discretion in emergencies.

This incident was just the beginning of my education in the complexities of police leadership. The discrepancy in how different officers were treated became even more apparent a few weeks later. A new blonde female officer, who was undeniably attractive, was dispatched to a domestic violence call at an apartment complex. When I arrived as her backup, I found her police car left in the travel lane, the driver's side door wide open, the engine running, and the shotgun lying exposed on the floor. She was already up on the third floor of the building.

I was about to secure her car when the sergeant and squad leader arrived. When I pointed out the multiple violations, they waved me off, saying, "You can just go back to your beat. We'll take care of this."

To my knowledge, she was never disciplined or even spoken to about leaving her car unsecured, running, and with a gun in plain sight. The double standard was glaring and disheartening.

These experiences taught me valuable lessons about leadership and the importance of fair, consistent rule-application. They showed me how poor management could create stress and potentially dangerous situations. As I progressed in my career, I vowed to remember these lessons, to lead with understanding and fairness, and to always consider the practical implications of the rules I enforced.

In your career, be prepared for supervisors who may be over-efficient but are not equal in how they apply discipline. You must realize who you're dealing with and how to ensure compliance. This lesson was driven home to me time and time again throughout my law enforcement career, starting with my experiences in Cary.

My subsequent encounter with another supervisor on that squad reminded me of how inexperience and a lack of proper leadership training can lead to misguided management. With no prior leadership experience, this newly appointed supervisor called me into his office one day. I had always prided myself on leading the department in statistics. I knew precisely what metrics were used to measure officer performance, and I consistently aimed to be at or near the top in all categories.

These metrics included the number of citations issued, arrests made, arrest warrants served, buildings checked, and DWI arrests. I've always believed that if I'm being paid to do a job, I should do it as thoroughly as possible throughout my entire shift. This work ethic resulted in me writing the most tickets, serving the most warrants, arresting the most drunk drivers, and conducting security checks on most buildings in the entire department.

On one night shift, I made two DWI arrests back-to-back. After processing the first offender at the Wake County Magistrate's Office and committing them to the Wake County Jail, I returned to Cary only to detect another drunk driver shortly after. I also arrested this individual and went through the entire process again, typically taking about two to two and a half hours.

To my surprise, I was summoned to the lieutenant's office. The department had grown, and the shift supervisor had been promoted to lieutenant, with a sergeant as an assistant. The lieutenant looked at me sternly and said, "You are not a highway patrolman. Stop making all of these DWI arrests."

I was taken aback. Never in my wildest dreams did I think anyone would discourage me from making DWI arrests. But I responded respectfully, "Yes, sir," and left his office, determined to find other ways to be productive.

On my next shift, I focused on serving warrants. Once again, I was called into the office. This time, the lieutenant said, "You're not a deputy sheriff. Stop serving all those papers."

Confused and somewhat frustrated, I tried to pivot once more. On the following night shift, I dedicated myself to checking the doors of businesses. I would call dispatch when I found an open door at a commercial establishment, request backup, and then search the building to ensure no one who shouldn't be was inside.

After several such checks and calls for assistance, I returned to the lieutenant's office. "You're not a security guard," he admonished. "Stop searching all those buildings."

At this point, I was at a loss. I couldn't arrest drunk drivers, serve warrants, or check businesses. So, I resorted to finding new construction sites with dedicated city streets and parked there to work on my homework for my associate's degree. When dispatch called for hourly safety checks, I would radio in with these obscure locations that no one had ever heard of because they were brand-new city streets.

Predictably, this led to another summons to the lieutenant's office. "I haven't seen or heard much from you," he remarked suspiciously.

"Just doing what you told me to do, sir," I replied, trying to mask my frustration.

This experience taught me about adapting to different leadership styles, even when they seem counterproductive. It also highlighted the importance of proper training for newly promoted supervisors. Without guidance on effective leadership and management techniques, even well-intentioned supervisors can create a demoralizing work environment.

As my career progressed, I encountered more situations that reinforced this lesson. When a full-time SWAT team opening became available (we called our SWAT team SOS for Special Operations Squad), I watched as another officer with prior experience but who had left the department was selected for the position. This was even though, according to the rules, he wasn't eligible because he hadn't been an employee for the required number of years. The department bent the rules, counting his previous experience and tenure to make him eligible.

This incident opened my eyes to the sometimes arbitrary nature of departmental decisions and the importance of networking and maintaining good relationships, even with former colleagues.

Feeling increasingly dissatisfied with my current position, I applied for a transfer to another squad. The lieutenant called me in, saying, "I don't want you to leave. I don't want you to have hard feelings, and this makes me look bad as a lieutenant if you want to leave." Reluctantly, I withdrew my application.

However, when another specialty position opened up, I applied again. This time, sensing my determination to leave, they transferred me to a different squad. It was the best thing that could have happened to me.

The new squad, under a different lieutenant, was everything I had envisioned full-time law enforcement to be. There was a true sense of brotherhood, where you want to spend time with your colleagues even on your days off. Several of us had motorcycles, and we'd often ride together. They even allowed me to purchase, train, and certify my police canine, K-9 Ricky. Working out at the same gym and hosting parties at each other's homes created the camaraderie I had always hoped for in law enforcement.

Under Lieutenant Kevin Kern's leadership, I flourished. Kern had experience and knew how to motivate his officers while maintaining discipline. This experience was quite different from my previous experiences and showed me the profound impact that good leadership can have on officer morale and performance.

A few months later, a new unit was formed - the directed patrol unit. This unit primarily dealt with traffic complaints but was also authorized to assist in investigations or any situation that required additional manpower. We weren't counted in the daily total of officers on regular patrol, which gave us more flexibility. Our primary focus was DWI enforcement and special events, but we could also back up other officers or assist with any task requiring extra hands.

On one bright summer day in 1989, a call came in - a silent alarm at a local bank in MacGregor Village. As I raced to the scene, my mind ran through all the possible scenarios. The situation became apparent when we arrived: there was no false alarm. The suspects had fled, leaving behind shaken employees and customers who had locked themselves inside the bank.

Recognizing the potential of the situation, I made a quick decision. "Lieutenant," I radioed, "permission to retrieve K-9 Ricky?" Ricky, my two-year-old German Shepherd, was a certified police canine. Still, due to department policies, he was housed at a local veterinarian's office rather than riding with me full-time.

With approval granted, I sped to Chatham Street Animal Hospital, collected Ricky from his kennel, and returned to the crime scene. As we began our track at the edge of the parking lot, I felt pride and excitement. This was what we had trained for.

Ricky's nose went to work immediately, casting back and forth until he locked onto a scent. We moved northward, crossing to the median of the shopping center's divided entryway. Suddenly, Ricky stopped, his attention fixed on something in the landscaping. There, partially hidden, was a torn piece of pantyhose—matching the description of the masks worn by the suspects.

Energized by this find, we pressed on. Ricky's training proved invaluable as he led us to another crucial piece of evidence - a nickel-plated semi-automatic pistol discarded in the pine straw. I called for an evidence tech, knowing that fingerprints could be essential in identifying our suspects. Each discovery brought us closer to them and reinforced the incredible value of K-9 units in law enforcement.

Our pursuit led us across US Highway 64 and into the MacGregor Downs subdivision. But as we turned down the first street, Ricky's trail went cold. Experience told me what this meant - our suspects had likely entered a vehicle and fled the area.

Thinking quickly, I approached a group of landscape workers tending to the subdivision's entrance. Their information proved crucial - they had seen two Hispanic males getting into a green Ford pickup truck and speeding away.

This detail was the missing piece we needed. Two hours later, one of our patrol officers spotted a vehicle matching the description. What followed was a textbook felony traffic stop. I was responsible for issuing commands to the suspects when I arrived as the second car on the scene.

The tension was tangible as we drew our guns and ordered the occupants out of the vehicle. The suspects complied, and we took them into custody without incident. A search of the vehicle revealed damning evidence: another pantyhose mask, a second semi-automatic pistol, and some of the stolen money.

Reflecting on this case and my early years with the Cary Police Department, I often think of how crucial teamwork and resources were to our success while serving. Those suspects might have slipped away without Ricky's tracking abilities, the observant landscape workers, or the coordinated effort of our entire department.

Over the years, as I continued my journey through the police department, I encountered supervisors with varying styles and priorities. Some, like the squad leader, seemed more focused on enforcing rules than on supporting their officers in the field. Others proved to be mentors, guiding me toward becoming a better officer and leader.

I employ you as you choose to serve others; remember that you can't do it alone. Be open to collaboration while using every resource at your disposal. Your partners - whether they're human colleagues or four-legged friends like Ricky - can be the key to turning a potential failure into a resounding success.

In my line of work, every day brings new challenges. But it's the willingness to work together and remain open-minded and resourceful that truly defines service excellence. So, remember to keep your eyes open, your mind sharp, and your heart open to help from unexpected places. That's how we make a real difference in the communities we serve.

Chapter 9

From Part-Time Police Officer to Police Chief (Part 3)

As my time in Cary, North Carolina, ended, I had no idea that the next steps in my career would lead to such transformative experiences. My work with the directed patrol unit deepened my understanding of community policing. Still, it was clear that an even more significant opportunity was just around the corner, one that would expand my skills and challenge me in ways I hadn't anticipated.

In 1990, that opportunity arrived when I was offered a position with the Wake County Sheriff's Office. The Drug Task Force had taken notice of my work, particularly with my K-9 partner, Ricky. By that point, Ricky had earned a reputation for sniffing out marijuana, cocaine, heroin, and various derivatives with impressive accuracy. If drugs were hidden nearby, you could count on Ricky to find them.

The man behind the offer was Sheriff John Baker, a towering figure physically and in terms of his legacy. Having spent time as an NFL player before leading the Sheriff's Office, he was known for his larger-than-life presence. But he also had a sharp instinct for identifying talent and recognized my work's potential. Sheriff Baker offered me a position on road patrol, a significant shift from my municipal policing days. It was a rare privilege to know your assignment before completing all your training, and the fact that I was one of the few to receive that kind of trust was both humbling and motivating.

Continuous Training

My new role placed me under the leadership of Captain Bill McClain on D-Platoon. It didn't take long to realize how different county law enforcement was from what I had experienced in Cary. The change in jurisdiction meant that we handled cases differently. One of the first calls I responded to was from a mother whose child had taken her car without permission. She expected me to file a report, which I would have done back in Cary. But this wasn't Cary, and the Sheriff's Office had a different protocol. When I checked with Captain McClain, his response was firm: "You don't take reports on those. Tell her to go to a magistrate. Let them decide if it needs a summons."

That lesson hit home. It wasn't just about what we did as officers and how we used our resources. At the Sheriff's Office, we were selective, focusing our efforts on more significant cases while leaving minor incidents to be handled by the courts. This required a mental shift but also helped me develop a stronger sense of prioritization—something that would become invaluable as I moved up the ranks.

Technology was also a big part of this transition. In 1990, the Wake County Sheriff's Office already used Tandy 100 computers—clunky, DOS-based laptops that allowed us to complete reports in the field. It might seem archaic now, but back then, it was revolutionary. The ability to complete reports electronically made our work more efficient and allowed us to focus more on what mattered—patrolling, investigating, and serving the community.

Adapting to this new environment also meant adjusting my approach to proactive law enforcement. One of the areas I was assigned to patrol was Holly Springs, where a bar—dubbed "Rock City"—had become a hotbed for crack cocaine activity. Policing a high-crime area like this required creative thinking, so I developed a strategy that used Ricky's presence to our advantage. I'd pull into the parking lot with Ricky barking, letting potential criminals know we were there, before surveying the area for signs of drug deals. Most of the time, the simple presence of my K-9 partner was enough to make people nervous, and I could conduct searches that often led to arrests.

But beyond these tactical moves, I made it a point to stay active in every aspect of my work. When things were quiet, I focused on serving civil papers—court documents like lawsuits or divorce papers that had to be delivered. At that time, patrol deputies in the county were responsible for serving these non-criminal court documents. I set daily goals: serve six papers, issue six citations, and make at least one arrest per shift. These self-imposed targets kept me productive and gave structure to my shifts. That kind of discipline became part of my work ethic, something I encouraged in others when I eventually stepped into leadership.

The environment at Wake County was different from what I'd experienced before in the best way possible. There was no micromanagement, no toxic leadership—just a team focused on doing the job right. That sense of professionalism and support allowed me to grow as an officer and a leader. I began formulating my ideas on leadership, combining what I had learned in Cary with the positive lessons I absorbed in Wake County.

I came to appreciate more than ever the value of continual learning. There's no room for stagnation in law enforcement. To serve effectively, you must keep growing, whether by embracing new technology, refining your communication skills, or learning from each new experience in the field. That mindset—that constant pursuit of improvement—was what I carried forward into every role I would later hold.

One summer afternoon in 1992, during an initiative known as *Operation Eagle*, I was reminded once again of the unpredictable nature of this job. We had set up alcohol-checking stations on roads near popular recreational areas, hoping to catch impaired drivers. I was stationed on Six Forks Road, not far from Raleigh, keeping an eye on the steady stream of traffic.

That's when I saw a bright yellow, immaculately restored 1970 Chevy Stepside pickup gleaming in the sun. The chrome wheels sparkled, and you could tell someone had poured a lot of love and effort into this vehicle. It stood out, sure, but something about it made me want to look closer.

As the vehicle approached and the driver realized there was a checking station ahead, they slammed on the brakes, squealing the tires, and abruptly made a U-turn before accelerating rapidly. It was an obvious attempt to avoid the checkpoint.

I sprinted to my patrol car, joined by one of the troopers manning the station, and we took off in pursuit. We chased the vehicle from our location near the lake to Highway 98, but the truck was nowhere in sight.

We turned around, realizing the only place it could have gone was into the public boat landing.

The Chevy pickup truck was there as we pulled into the parking lot. The engine was still running while the driver's door was opened wide. Without hesitation, I got K-9 Ricky out and put his tracking line on him. The trooper, eyeing Ricky skeptically, asked, "Is that dog any good?"

I couldn't help but smile. "If you can keep up with us, you'll find out," I replied. I knew from experience that most people who tried to follow us couldn't match Ricky's pace through the woods. They were often more concerned about their uniforms than catching the suspect. On the other hand, I was focused solely on apprehending the person we were chasing.

The trooper, clearly intrigued, made a wager. "If the dog finds him, I'll buy him a steak dinner tonight," he declared. I accepted the challenge, and Ricky and I took off on the track.

It wasn't long before the trooper fell behind and was no longer in sight. Ricky and I continued on the trail, his nose to the ground, following the scent. As we progressed, Ricky started to pull harder on the tracking line. I watched his ears intently – I had learned to use them like a gunsight. Wherever they were pointing, the suspect would usually be right in the middle, just as if I were aiming a firearm.

Suddenly, Ricky's ears perked up, and he began pulling harder on the tracking line. He started to bark and growl, a clear sign that we were close. As we approached a fallen tree, I could see someone hiding behind it, lying flat on the ground.

I drew my weapon and began issuing commands. "Put your hands above your head where I can see them," I shouted. "Slowly stand up. I'm warning you – the dog will bite, and I will shoot if you don't comply."

The suspect slowly rose, hands in the air, and I placed him in custody. Out of habit, I asked, "Why did you run?" His answer was a stark reminder of the real-world consequences of our work. He explained that he worked for the power company and that a DWI conviction would cost him his job.

But our work wasn't done. Ricky continued pulling on his harness, indicating someone else was in the area. I released his leash and let him go. Shortly after, I heard him barking and growling again. This time, I saw another man standing behind a tree, trying to hide.

By this time, the trooper had finally caught up with us. He drew his weapon as I gave Ricky the command to go down. The trooper then approached and handcuffed the second suspect.

Incredibly, Ricky alerted once more. He jumped up and took off, leading us to a third subject hiding in the woods. Thanks to Ricky's keen nose and training, we had apprehended three suspects who would have otherwise escaped.

We later learned that these were the passengers from the truck. Initially, we couldn't determine how many people were in the vehicle other than the driver. When we arrived at the parking lot by the boat ramp, only the driver's side door had been open, not indicating additional suspects.

The trooper took custody of the driver and charged him with DWI. When he returned from the magistrate's office, he approached me with a sheepish grin. "Do I really have to buy the dog a steak dinner?" he asked.

I couldn't help but chuckle. "I think he'll settle for a hamburger," I replied. "He's not that discerning." True to his word, the trooper went through a drive-thru and bought Ricky a hamburger, which my faithful partner thoroughly enjoyed.

This incident was yet another testament to the invaluable role that K-9 units play in law enforcement. They help keep officers safe and are incredibly effective at locating suspects who attempt to flee and hide. Experiences like these reinforced my belief in the importance of K-9 units and would influence my decisions as I moved into leadership roles.

As winter approached in 1993, my expertise with K-9 Ricky was again called upon. The drug task force asked me to accompany them on the execution of a search warrant for a known crack dealer. The target was a mobile home in a local trailer park.

I always advised the agents and officers I worked with to allow the dog to do the initial sweep for drugs once the scene was secure. My experience had shown that if deputies went in first, they often disturbed the scene, scattering evidence and making the dog's job more difficult. It was much more efficient and productive to let the K-9 do the initial search.

The team had learned to trust this process after several instances where Ricky found drugs that the deputies had missed in their initial searches. On this occasion, once the occupants were secured, Ricky and I entered the mobile home alone.

The interior of the home was in complete disarray. Moldy dishes filled the sink, and debris was scattered everywhere. As we searched, Ricky indicated a laundry basket overflowing with clothes. He pushed it over and started rooting through the pile of dirty laundry, eventually pulling out a bulging Crown Royal bag.

I opened the bag from him, revealing a substantial amount of U.S. currency. The final count came to over $30,000. This was a significant find, as at the time, cocaine was prevalent, and many users would roll up dollar bills to snort the drug. We were often told that almost all money in circulation would have some cocaine residue on it due to this practice.

This moment was a test of integrity. We did locate some drugs during the search, but the cash was undoubtedly the highlight. If I had been dishonest, I could have easily slipped that bag into my cargo pocket and walked out without anyone knowing the difference. But that thought never crossed my mind. I believe that integrity is defined by what you do when no one looks, not just when you're being observed.

This search was another successful conclusion to a drug investigation, resulting in the arrest of a crack cocaine dealer and the seizure of a substantial amount of money that could be used to further other investigations and combat the spread of this dangerous and highly addictive drug.

As I approached my eighth year in full-time law enforcement, I realized that police departments could improve training, community service, and overall operations. However, I also recognized that only someone in the role of chief of police would have the authority and ability to implement such changes.

At the time, I was working on my two-year degree at Wake Technical College. Due to my rotating shifts and the inability to take leave to attend classes, I could only take one class at a time. This meant that it took me almost eight years to complete what was typically a two-year degree. But I persevered, consistently working at it and enjoying the classes despite the slow progress.

In 1994, as I was nearing completion of my degree, an opportunity presented itself that would change the course of my career. I was offered the position of police chief in Bladenboro, North Carolina.

Chapter 10

The Bladenboro Encounter

Bladenboro was a small rural community in the southeast portion of the state, not far from Myrtle Beach, South Carolina. The Bladenboro Police Department consisted of four full-time police officers and a chief. This meant one officer per shift, with the chief overseeing operations and serving as backup when needed. The chief also had to fill in for vacations, training, or sick leave when officers couldn't cover their regular schedules.

I had purchased K-9 Ricky with my own money and trained him using my funds, so naturally, he came with me to this new role as chief of police. His skills would prove invaluable in this new setting.

Upon arriving in Bladenboro, I discovered that the local sheriff's office also had a drug task force. I offered them the services of K-9 Ricky, which they gladly accepted. The head of the unit, Phillip Little, was particularly interested in my suggestion of allowing the dog to go first in searches.

Our first opportunity to implement this strategy came soon after. We were called to assist with a search of a small compound consisting of a mobile home, several outbuildings, and a larger maintenance shop-type building, all enclosed within a perimeter.

The deputies secured the subjects and searched all the buildings, including the residence. They went through every drawer, looked behind furniture and pictures - anywhere they thought someone might hide drugs. However, they didn't give Ricky a chance to search.

After they had completed their search and left the buildings, I took Ricky into the mobile home. Almost immediately, he found some crack cocaine wedged between the mattress and box springs of one of the beds – something the deputies had missed.

We then moved to the maintenance building. As I conducted a perimeter search of the interior, Ricky quickly located two large bags of marijuana. The drug dealers had cleverly hidden them between the mower deck and the blade of a riding lawnmower. The blade was holding the marijuana up against the deck of the mower. Again, the deputies overlooked this, never considering turning over the vehicle and looking underneath.

This experience powerfully demonstrated Ricky's abilities and the effectiveness of our search strategy. The deputies quickly acknowledged that I was right, and from that point on, Ricky was always the first to go in and find hidden drugs.

When stepping into a leadership role, especially in a small community like Bladenboro, one of the biggest challenges can be navigating the dynamics of working with others who have already established methods and routines. It's common to encounter resistance, especially when introducing new ideas or approaches.

As a law enforcement officer or in any service-oriented role, you may find that your expertise isn't always immediately welcomed, and your suggestions might be brushed aside in favor of the tried and tested ways others are more comfortable with. In these situations, it's essential to remain patient and persistent. Everyone in the field is working toward the same goal—ensuring the safety and well-being of the community—but different levels of experience and perspectives can sometimes cause friction.

The key to success is demonstrating your expertise and learning to blend it with the skills and knowledge of others. That means being willing to let others take the lead at times but also knowing when to step up and show the value of your approach when the opportunity presents itself.

Another critical aspect of working in environments where your ideas may initially be ignored is maintaining professionalism and respect for others, even when disagreements arise. Instead of becoming frustrated or discouraged, take those moments as opportunities to prove the effectiveness of your methods through action, as I did with K-9 Ricky.

When others see tangible results, they are far more likely to respect your input and welcome future collaboration. By focusing on shared objectives and respecting the contributions of all team members, you can create an environment of mutual respect and trust where everyone feels valued and the best ideas—no matter where they come from—rise to the top.

As you seek opportunities to serve others, remember that contributions don't always come from the same team or department in any service field. Sometimes, the expertise needed lies with individuals or groups outside your immediate circle, and recognizing when to involve them is the key to success. No one person or team can be expected to have all the answers, especially when complex situations arise that require specialized knowledge.

The ability to reach across departments, agencies, or even levels of government service to bring fresh perspectives can make all the difference in solving a problem. When we collaborate beyond our teams, we open the door to creative solutions and ensure no stone is left unturned.

This proved true for me during a critical investigation when I called upon the State Bureau of Investigation (SBI) to help with a case beyond the scope of our small-town department.

It happened late one night when I was abruptly awakened by a call informing me of a robbery at our local IGA grocery store. I immediately responded to the scene, where I found the front glass window shattered and several visibly shaken employees inside. As I began investigating, I separated the employees to get independent accounts of what had transpired.

The store manager was the first person I interviewed. He recounted a chilling series of events after the store had closed for the night. As he got a drink from the fountain near the restrooms, two armed black males emerged—one from the women's restroom and another from the men's. They took him at gunpoint to the hardware aisle, where they bound and gagged him with duct tape and telephone extension cords.

The suspects then forced the manager to use the PA system to call all employees to the meat-cutting room for a meeting. When the unsuspecting employees arrived, the armed men emerged from hiding and took everyone hostage. What followed was a horrifying ordeal for the employees. The suspects went so far as to force a female employee to strip naked and subjected her to degrading treatment with their weapons.

After terrorizing the staff, the robbers took the manager to the front of the store and had him open the safe. They cleared out all the money and grabbed several cartons of cigarettes before fleeing with the manager's keys, leaving all the employees locked inside the store.

In a display of quick thinking and bravery, one of the female employees managed to free herself. She ran to the front of the store, grabbed a large ashtray, and smashed out the front window to escape. Once she reached her nearby home, she immediately called the police, which led to our response.

As we gathered more details about the suspect's appearance, weapons, and equipment—including walkie-talkies and police scanners—I realized this case was far beyond the capabilities of our small, five-person department. I decided to call the North Carolina State Bureau of Investigation (SBI) for assistance.

When two SBI agents arrived about an hour later, I briefed them on what we had learned. A breakthrough emerged as we ran records checks on individuals matching the suspects' descriptions. One of the suspects, with two filled-in teardrops by his right eye and the name "LaLola" tattooed on his neck, matched the description of individuals involved in a string of Pizza Hut robberies in the Fayetteville area.

The modus operandi (MO) of these Pizza Hut robberies—calling in an order just before closing, entering the store, and then taking employees hostage—bore striking similarities to our grocery store incident. This connection provided a crucial lead that would prove invaluable in our investigation.

What followed was an intense 48-hour period of non-stop work. My team and the SBI agents worked tirelessly, pushing through exhaustion, knowing we were gaining ground with each piece of information we uncovered. Our investigation led us to involve the Lumberton Police Department and their Drug Investigation Unit, who had information about suspects in a check-writing scheme operating out of a local hotel.

The description of one of the check-writing suspects, with "LaLola" tattooed on his neck, matched our grocery store robber. This connection allowed us to locate the suspects in a motel, where we executed a search warrant and made multiple arrests.

Some evidence discovered during the search includes a stolen Mustang used as a getaway car in the grocery store heist, a stolen check writing machine used in the fraudulent check scheme, walkie-talkies taken from a nearby hotel, police scanners bought from Radio Shack and programmed with local law enforcement frequencies, and the guns used in the robbery.

In the end, we were able to charge the suspects with two counts of armed robbery and eight counts of kidnapping for the Bladenboro incident alone. This case demonstrated that serious crimes can occur even in small towns like Bladenboro, and it's crucial to have solid interagency cooperation to solve them effectively.

Remember that service isn't always about what you can accomplish alone; it's about how well you collaborate with others to achieve a common goal. No matter how capable or experienced you may be, there will come a time when the resources and perspectives of others are essential to fulfilling your role effectively. This case with the grocery store robbery underscored that in service—whether to your community, your team, or even yourself—you sometimes need to step back and recognize when it's time to bring in outside help.

This often demands humility. It requires acknowledging that other agencies, departments, or even individuals outside your usual circle may be vital to solving a problem or making a significant impact. The collaboration between my department, the SBI, and even the Lumberton Police Department is a perfect example of how embracing interagency teamwork allowed us to overcome limitations and achieve something we couldn't have accomplished on our own. Ultimately, service is about offering expertise and amplifying your community's strength by embracing others' contributions.

This experience reminds me that in service, the shared commitment to the greater good ultimately leads to success. Whether in law enforcement, healthcare, or any other field, genuine service requires recognizing when to take the lead, step aside, and call on others to help meet the needs of those we serve.

As I continued as chief of the Bladenboro Police Department, I was later introduced to some unique collaborations in law enforcement. Many people may not be aware that the National Guard has a counter-drug unit. One of the missions of this unit is to assist local law enforcement with their drug investigations using military equipment.

In late 1994, the National Guard approached me to participate in Operation Blade Runner. This operation would use military personnel and Army helicopters to search for outdoor marijuana grows and then direct police officers to where the drugs were located. We were also able to use some of the soldiers for surveillance operations.

One afternoon, while I was working with the drug unit, we received a call from a helicopter stating that there was a large hydroponic grow operation in Crawley Swamp. The helicopter crew reported seeing several raft-like structures floating in the swamp.

The helicopter guided our team through the woods and into the swamp area. The water was about waist-high as we waded in. What we discovered was a piece of evidence of human ingenuity, albeit applied to illegal activities. The suspect had created makeshift rafts using three plastic five-gallon buckets, building a frame of two-by-fours to hold them together. They had drilled holes in the bottom of the buckets so that water from the swamp could be drawn up into the bucket to water the marijuana plants. These rafts were then floated out and secured with ropes to trees to prevent them from drifting away.

What we stumbled upon was no small operation. It took me and several deputies two full days to harvest these plants. Once they matured and were harvested for illegal use, their street value was estimated to be well over several million dollars. The ingenuity of the operation didn't stop at the rafts. The suspect had also employed another clever technique. They had taken some of the five-gallon buckets, tied them to saplings, and suspended them from the trees over the swamp. A wick system was set up to draw water from the swamp through the wick and into the plant. It was clear that whoever was behind this operation had put much thought and effort into their illicit enterprise.

What struck me most about this setup was the level of precision and care that had gone into it. This was not a hastily thrown-together operation. The person responsible had done a remarkable job in construction and camouflage. All the rafts were neatly made, the buckets were painted gray to blend in with the murky swamp water, and the wood was painted brown to match the surrounding trees. This wasn't a quick, shoddy operation but something the cultivator had intended to last for multiple growing seasons.

Despite the operation's sophistication, we could never catch anyone in the act. After our first day of harvesting, we utilized the National Guard soldiers to survey the area. Some of their soldiers were left in the swamp to act as scouts, tasked with notifying us if anybody came to tend to the marijuana grow or to harvest the plants we hadn't yet removed. Unfortunately, no one ever returned to the swamp during our surveillance period.

We found an ancient, single-wide mobile home near the swamp where the plants were located. It appeared to be set up as a base of operations for whoever was running the grow. On the trailer's roof was a 55-gallon bucket, clearly used to gather rainwater—a hose led from this makeshift water collection system into the trailer bathroom. An old Army generator supplied power.

Despite these signs of habitation, there was no concrete evidence to determine who had been staying in the trailer. Someone had slept and eaten there, but it didn't appear to be daily. We found no personal effects or other clues that could lead us to identify the owner or occupant.

Even after several more days of surveillance, no one ever returned to the trailer or the swamp. While we weren't able to make an arrest, the operation certainly made a significant dent in the availability of marijuana in the area that year. We continued to monitor the location to see if anyone would ever return to try and reestablish their hydroponic growth. Still, the sophisticated operation appeared to have been abandoned for good.

Criminal businesses are constantly evolving, and it is crucial for law enforcement officers always to remain one step ahead. This can be achieved by emphasizing and implementing initiatives that value interagency cooperation. Without the support of the National Guard and their equipment, we could never have discovered that hidden operation in the swamp.

During those earlier days, as I settled into my role as chief, I began to implement some of the changes I had long believed were necessary. One of my first initiatives was to institute a reserve officer program. This was a strategic move to augment our full-time staff, allowing us to have more than one officer on the street at a time. It was a lesson in resource management – how to maximize our impact with limited means.

Also, I quickly realized that being a supervisor came with its own unique set of challenges. The idealistic views I had held as a patrol officer or deputy sheriff about how things should be run were now put to the test. The realities of budgets, politics, and personnel management were far more complex than I had imagined.

One of my long-held beliefs was the importance of regular firearms training. As an instructor, I was acutely aware of the national statistics on police accuracy in real-world situations. The fact that officers only hit their intended targets 20% of the time in actual street encounters was deeply concerning to me. This meant that eight out of ten bullets fired by police officers were missing their targets, potentially endangering innocent bystanders, damaging property, or worse.

I had always dreamed of implementing quarterly firearms qualifications, going above and beyond the state requirements. This additional training would significantly improve our officers' proficiency and, by extension, public safety. However, as I dove into the department's finances, the harsh reality of our budget constraints quickly became apparent.

My first board meeting was an eye-opener. I had barely had time to review our financial situation, but what I found was alarming. By February of each year, the department had been operating in the red for the past five years. The primary culprit? Vehicle repairs. When we got them, the department had been buying used highway patrol cars, which were already well-worn. With no warranty and frequent major mechanical failures - engines, transmissions, differentials - our repair costs were astronomical.

Armed with this knowledge, I went into the meeting with a plan. I proposed a leasing program for our police vehicles. In my mind, it was a perfect solution. The lease payments wouldn't cost us any more than we were already spending on buying used cars and their subsequent repairs. We could trade in the cars every three years for new models, keeping our expenses stable and predictable. Plus, our officers would have new, reliable equipment, presenting a more professional image for the department.

I stood before the board, confident in my proposal, only to be summarily shot down. One board member's comment particularly stung: "That's because you're wasting the money."

Taken aback, I responded, "Sir, I haven't spent a penny since I've taken this job. You'll notice I'm not wearing a uniform. That's because we don't have any money to buy a uniform. You'll notice I don't have a badge because your last chief stole the badge."

The board members exchanged uneasy glances as the reality of our situation sank in. I pressed on, recommending they increase the department's budget by at least $25,000 for the next fiscal year to offset these recurring costs.

This experience was a harsh lesson in dealing with city commissioners who often lacked education, business experience, or government training. Yet, these individuals made crucial decisions for their departments and the community.

I learned quickly that to be effective, I needed always to have a well-thought-out plan and several alternative courses of action. It became clear that explaining the importance and benefits of each proposal to the community was crucial.

Another significant challenge I faced was related to the "Click It or Ticket" program, a grant-funded overtime enforcement initiative from the North Carolina Governor's Highway Safety Program. When I first arrived in Bladenboro in 1994, I noticed that only about 25% of motorists were using their seat belts despite the occupant restraint law being in effect since 1986. The crash data showed an increasing number of accidents yearly, with injuries outnumbering the crashes.

Drawing on my connections from my time with the Wake County Sheriff's Office, I applied and received a grant for the "Click It or Ticket" program. My approach was multifaceted: first, I launched an education campaign, visiting manufacturing companies and other large employers to explain the importance of seat belt use. I also spoke at the high school, addressing seniors and juniors, and became a regular speaker in driver's education courses.

After the education phase, we set up checkpoints for seat belt and child restraint violations. The campaign was intense—for the first time in Bladenboro's history, we had five police cars on duty at once, all focusing on seat belt enforcement.

The results were remarkable. Within 12 months, seat belt use in Bladenboro reached 100%, and we had drastically reduced the number of injury accidents to just one - involving a speeding, unbelted high school driver. This success led to several awards, including the State Officer of the Year Award from the Veterans of Foreign Wars, a certificate from the Governor's Highway Safety Office, and a trophy from the International Association of Chiefs of Police.

However, this success came with its own set of challenges. The commissioners instructed me to return the grant money at a subsequent board meeting. They were receiving complaints from their neighbors about the increased number of tickets being issued. I stood my ground, explaining that we had already agreed to accept the money. Moreover, we started using it and were obligated to continue fulfilling the agreement they had signed with the state.

These experiences taught me valuable lessons about the complexities of police administration. There were stresses and considerations at the administrative level that weren't always obvious to those on the outside. Budgets, grants, and problem-solving – all play a crucial role in running an effective department.

In any leadership position, especially in public service, there will be moments where your decisions, no matter how successful, are met with resistance or pushback. Sometimes, the community you're trying to serve doesn't fully understand or appreciate the broader impact of your implementing changes. It's a lesson in navigating the balance between what's popular and necessary for long-term improvement.

When serving in any capacity, it's crucial to recognize that administration goes beyond just leading people or making immediate decisions. There's an entire world of budgeting, grant writing, and managing relationships with various stakeholders that many may not see. It's easy to assume that success will always be celebrated, but real-world service often requires defending your methods and standing firm in your mission, even when faced with opposition.

When stepping into leadership, particularly within public service, your understanding of these administrative aspects early on will be invaluable. It's not just about being on the front lines; it's about knowing how to secure the resources your team needs and working within the constraints of your environment. This often means building relationships with people who work behind the scenes—people in administration who understand how to navigate budgets and secure grants.

By asking the right questions and seeking out those connections, you prepare yourself to serve and lead effectively. Part of that preparation is realizing that when you're in a position where you're responsible for others, there are often difficult decisions to make. Not everyone will be happy with the results. But if your focus remains on the greater good and the long-term benefits for the community, you'll find that the challenges of administration can be met with patience and persistence.

Service leadership is about both vision and practicality. It's about finding ways to push your department or organization forward despite scarce resources. Whether finding the right grants to fund essential programs or making tough calls in the face of opposition, these experiences teach us that true service requires passion, strategy, resilience, and a willingness to advocate for what's right—even when it's not easy.

Given Bladenboro's limited tax base, I knew we would never have an abundance of brand-new equipment. To overcome this challenge, I had to become adept at finding grant sources to either secure equipment directly or obtain funding for it. Through persistent effort and strategic thinking, I increased the professionalism of each municipal agency I served in North Carolina.

After more than two years as the chief in Bladenboro, dealing with an often tumultuous town council, I was offered a new opportunity – the position of Chief of Police in Norwood, North Carolina. Located in Stanley County, about 30 miles east of Charlotte, Norwood presented a new set of challenges and opportunities to my journey in law enforcement.

Chapter 11

Norwood's Chief of Police

After the lessons learned from Bladenboro's board, I knew better than to rush into my next role without some due diligence. When the opportunity to serve as Chief of Police in Norwood presented itself, I attended one of their town board meetings before accepting the job. It wasn't just a matter of formality but about understanding the culture I would enter.

To my relief, the environment was worlds apart from the dysfunction I had dealt with before. The commissioners in Norwood worked cohesively, handling town business gracefully and efficiently. Meetings ran on schedule and ended with a rare community spirit—sometimes, there was cake, pie, and ice cream. It was clear that Norwood wasn't just another job—it was the kind of place where I could make a difference without constant internal battles.

Encouraged by the board's professionalism, I accepted the position. The police department was small but well-organized: seven people—a chief, a captain, a sergeant, four patrol officers, and a part-time secretary. It was manageable, but I knew from experience that even a small department could present challenges.

One of the trickiest aspects of leadership is maintaining a balance between professional authority and personal rapport. Getting too close to your subordinates could make it difficult when tough decisions or disciplinary actions are needed. That lesson returned to me shortly after I took the position, particularly with one incredibly welcoming officer.

Roy Taylor PhD

He lived on Lake Tillery and had a picturesque lake house with a boat dock. During those early months, I lived alone in Norwood—my wife was back in Bladenboro, waiting for our home to sell. The loneliness of those days made the officer's invitations for boat rides and dinners by the lake incredibly tempting, and I accepted. It seemed harmless then, but I failed to consider how it might blur the lines between friendship and professional duty.

Months later, as I was cleaning out the desk of the previous chief, I stumbled upon a receipt for two handgun magazines for a Sig Sauer pistol. That small piece of paper triggered a memory. The same officer who had welcomed me so warmly had once told me a story about when he was hired. The department, lacking enough equipment due to budget constraints, had asked him to use his gun temporarily until a new one could be purchased. He only had one magazine, so the department agreed to buy two extras for him. It was understandable, given that he had been hired through a COPS

FAST grant from the Department of Justice, which provided funds for additional officers but not always for equipment.

Armed with this information, I approached the officer and asked if he still had those magazines. I assumed he would understand the question was procedural, but his response was immediate and defensive: "They were mine," he insisted.

I corrected him calmly but firmly. "No, the department purchased them for you. So you can either return or pay for them, but it's department property, and you can't sell it."

What happened next took me by surprise. He became enraged, shouting about how he thought we were friends, and then he said something cryptic that made my stomach drop: "At least I don't steal guns."

Stunned, I pressed him for more details. At first, he tried to backtrack, but I knew better than to let this go. This conversation had just turned into something much more severe than missing magazines. I started recording as he unraveled a story that would shake the department's foundation.

He told me that the captain had stolen a gun from a Hispanic male arrested for DWI. According to the officer, after the man was arrested, his car had been left on the side of the road. The captain had later asked the officer to retrieve a Colt Woodsman pistol from under the car seat—a collector's item, no less. The officer described how he went to the county jail, retrieved the car keys from the jailer, drove back to the car, took the gun, locked the vehicle, and returned the keys—all without ever telling the prisoner about the gun.

When I asked what had become of the gun, the officer claimed he had placed it on the chief's desk, as was supposedly customary. But then said he saw the captain using that same gun during their last annual firearms qualification. The story had holes—officers were supposed to keep their evidence, not leave it lying on the chief's desk. Still, it was severe enough that I had to investigate further.

I called the captain in for questioning and advising him of his Garrity rights, which are similar to Miranda rights but specific to non-criminal investigations in law enforcement. Under Garrity, he was compelled to answer truthfully or face termination, though his statements couldn't be used against him in a criminal proceeding.

The captain admitted leaving the gun in the car and asking the officer to retrieve it. He also confessed to taking possession of the firearm. But when I asked him where the gun was now, his answer dropped like a bomb—he had sold it during his divorce, along with several other personal firearms. This was no minor infraction. I had to suspend him without pay and begin the termination process. I also notified the Criminal Justice Training and Standards Commission to revoke his certification.

The officer who had exposed this mess wasn't off the hook either. While he had provided valuable information, he had violated department procedures and acted inappropriately. His defiance and lack of accountability left me no choice but to suspend him, notifying him that termination was imminent.

Both terminations were difficult but necessary for the department's integrity—there was no room for corruption or negligence, and it was not under my watch. I sent the paperwork to the Training and Standards Commission for both men, requesting that their certifications be revoked. The officer, however, lawyered up and negotiated a deal where he could resign in exchange for a neutral reference. The commission decided not to revoke his certification, considering he had followed the captain's orders. The captain, however, was done with law enforcement for good.

This situation reaffirmed an essential truth about leadership: you cannot allow personal relationships to compromise your judgment. The officer's belief that we were friends clouded his understanding of the role I played as his superior. It was a hard lesson for both of us but an essential one. See, the decisions you make as a leader affect your immediate team and the broader community you serve.

It's challenging enough to grapple with the idea of someone losing their job and the financial ramifications that will have on their life. In law enforcement, the consequences of termination are even more severe. Unlike many other professions, you can't quit one agency today and get another job tomorrow. The requirements for certification in North Carolina require a new background investigation every time an officer moves to a new department.

Let me break this down for you: once you're a certified officer in North Carolina, it takes at least 90 days to process a package that includes a physical, drug screen, and a complete background investigation. This package then gets submitted to the state for another review before the officer is awarded certification and the paperwork to swear them in. Any negative information about why an officer left their previous employment would be validated by the Training and Standards Commission to ensure we weren't passing problematic officers to other agencies.

So, when making these decisions, I had to consider not only that I was taking away their livelihood today with their full-time job but potentially impacting their ability to work in law enforcement for the rest of their lives. Depending on the circumstances, if they appealed their termination, the Training and Standards Commission might rule for a 12-month suspension, a five-year suspension, or even permanent revocation of their certification.

Understanding these ramifications is crucial for a leader. None of these decisions should be taken lightly or made in the heat of the moment. They always require reflection and a thorough examination of the facts, followed by a cool, calm decision based on policies, procedures, and the law.

During my time in Norwood, after dealing with the corruption within the department, I knew that rebuilding trust—internally and with the community—would be paramount. The incidents with the captain and the officer had shaken the foundation of our small force, but they also presented an opportunity for genuine change.

My experiences in Bladenboro had taught me the importance of community engagement, and I was determined to implement those lessons in Norwood. The first step was to assess the community's perception of our department. What I found was disheartening but unsurprising: many viewed us as incompetent, likening us to a bunch of Barney Fife's characters from the hit 1960s show "The Andy Griffith Show." We needed to bridge the gap between perception and reality.

As I continued my assessment, I realized the biggest challenge was reshaping how our officers engaged with the people of Norwood. Our officers were hardworking and dedicated, but they had fallen into a pattern that separated them from those they were meant to protect. Our patrol cars had become a physical and symbolic barrier between them and the community. There was little to no interaction outside of arrests, traffic stops, or moments of crisis.

I knew something had to change. Trust and open communication were essential for a department to serve its community effectively; it wasn't enough for us to uphold the law; we had to become a part of the town's fabric. I decided the solution lay in community policing—a philosophy emphasizing building relationships rather than walls. But as with any systemic change, there was resistance. Some officers were obstinate, clinging to old ways out of fear or a lack of understanding. I knew that pushing this transformation was non-negotiable.

"This is like a train," I told them. "You can't stop it. You either get on board, or you'll get run over." My conviction in this matter was clear.

This was not just about doing our jobs differently—it was about saving the future of our town's safety and restoring faith in the police.

Once I'd completed the initial phase of training, we needed to hear directly from the residents. I devised a community survey and had the officers deliver it door-to-door themselves. It wasn't just about the questions on the form. It was about showing up at someone's door without a uniform agenda—just an open hand asking for feedback.

The survey revealed that while Norwood was generally safe, many older residents felt uneasy due to quality-of-life issues—teenagers hanging out at car washes or playing loud music at convenience stores. These weren't criminal concerns but real worries for those living there.

During this time, I divided the town into seven zones, assigning an officer to each, and organized community meetings where we would sit down face-to-face with residents to talk about their concerns. Every officer was issued a pager, and we ensured people knew they could reach out to their designated officer whenever they had non-emergency questions or issues.

Having one familiar face to go to, someone they felt connected with, worked wonders for the community's trust in us. Suddenly, we weren't just an impersonal force looming over them; we were neighbors, friends, people they could rely on.

This was not the kind of change that could happen overnight. It took patience, consistency, and humility. But as the months went by, I began to notice the shift. The town that had once viewed us skeptically now saw us as partners in keeping Norwood safe.

While community policing strengthened relationships, we still had to face the darker side of law enforcement: drugs, prostitution, and liquor houses. These weren't just local issues; they extended across county lines. Years before, there had been a task force called the USA Drug and

Violent Crime Task Force that dealt with these problems, but it had disbanded due to funding issues.

I contacted the surrounding agencies and convinced them to revive the task force, albeit without any external funding. We agreed to pool our resources, working across jurisdictions to combat everything from drug trafficking to cockfighting. The officers, particularly mine, found this collaborative effort exhilarating. It gave them a sense of purpose and the opportunity to engage in operations beyond the usual routine.

But no matter how carefully we planned, danger was always lurking. On one such operation, after gathering enough evidence through controlled drug buys, we raided a house suspected of trafficking large quantities of marijuana. The moment we knocked on the door and announced our search warrant, gunfire erupted from the second floor. It was one of those moments where everything you've trained for is put to the test.

Amid the chaos, my officers acted swiftly, knocking out floodlights to gain the upper hand. After forcing our way inside, we secured the lower part of the house and discovered that our suspects had jumped from the second floor to escape. Despite the gunfire and the escape, we managed to arrest two people and seize the bulk of their drug operation. As it turned out, one of the individuals involved was on the FBI's 10 Most Wanted list for murder.

In small towns like Norwood, you don't expect to come face-to-face with someone so dangerous. Yet, here we were. It was a reminder that criminal activity doesn't respect borders or town sizes. It was a victory for our department and the town itself.

A year after we implemented community policing, the transformation was undeniable. The bond between the police department and the community had strengthened. People started to take pride in their police force, and we made it a point to highlight our successes in the local newspaper. My goal was simple: for every positive story about the police in the paper, the community would see that we were more than enforcers; we were protectors, partners, and problem solvers.

This commitment to building trust eventually caught the attention of the National League of Municipalities. They sent assessors from Washington, D.C., to evaluate our program for a national award in community policing. I was both proud and anxious when they arrived, but the following moments were some of the most gratifying of my career.

We started the tour at the local elementary school, where my Adopt-A-Cop program was in full swing. Many of my officers initially grumbled about playing with kids and eating lunch in the cafeteria. But over time, even the most hardened officers softened. They began to enjoy it, finding meaning in their connections with the children.

From there, I took the assessors to meet one of our most unexpected allies—a local prostitute. Months earlier, I had responded to a domestic disturbance at her residence and had helped her resolve the situation peacefully without arrests. This act of kindness and understanding had turned her into one of our best confidential informants. She later began providing me invaluable information on drug dealers and liquor houses. When she spoke to the assessors, she was unreserved in her praise for how the police had changed since I took over. She knew the streets better than anyone, and her testimony proved our efforts worked.

Ultimately, Norwood won first place for its community policing program, and I received an award from the Governor's Crime Commission for my efforts. But the valid reward was something much deeper—the knowledge that we had made a difference, not just in lowering crime but transforming how the town saw us and how we saw ourselves. The lessons we learned in Norwood continue to guide me to this day: leadership is about service, and trust is something you have to earn daily.

Norwood was my training ground in many ways, sharpening my skills as a leader who could navigate complex personnel issues while maintaining the town's and its citizens' trust. My job wasn't just about catching criminals or enforcing laws—it was about instilling confidence in the badge and the justice system itself. Every decision, whether disciplinary or commendatory, reflected that responsibility.

Even as I worked through the challenges of running the department, my mind often drifted to the bigger picture of leadership in law enforcement. I thought about how much I had learned over the years and how each town and each position had taught me something new. From Cary to Wake County, Bladenboro to Norwood, I constantly refined my understanding of serving.

And that's where I found my footing—not just in the day-to-day duties of a police chief, but in the life of service itself. It became clear to me that leadership wasn't static. You had to keep evolving, learning, and growing. Mistakes would be made, but how you responded to them defined your legacy.

Reflecting on my time in Norwood, I also began to look ahead. This job prepared me in more ways than I anticipated. It solidified my understanding of community policing—where trust, communication, and integrity were at the forefront of everything we did. I knew I would carry these lessons forward as I moved on to my next role.

The opportunity in Liberty, North Carolina, came shortly after. A town larger than Norwood, with a police department that consisted of thirteen officers, Liberty was a step up. However, I made a crucial error before accepting the position—I didn't attend a council meeting or dig deeply into the department's history. It wasn't until I had taken the job that I learned Liberty had gone through eight chiefs in five years. The churn was a red flag I should have seen, but I believed that I could be the one to bring stability to the department.

Looking back, I realize that the move to Liberty wasn't without its challenges, but it continued my growth journey. Leadership in law enforcement isn't just about managing people or ensuring policies are followed—it's about serving your community with unwavering commitment, no matter the obstacles.

As I settled into my new role at Liberty, I couldn't help but think back on my journey. From a part-time officer to the chief of a department larger than the national average, each experience had been a learning curve. The decisions I had made, the mistakes, the victories—they all shaped who I had become as a leader. Every day in law enforcement taught me something new, and I was grateful for the lessons, however tough they were.

Chapter 12

Navigating Choppy Waters

Have you ever found yourself in a situation where doing the right thing while serving felt like walking a tightrope over a pit of hungry alligators? My time as the police chief for Liberty, North Carolina, was a crash course in navigating treacherous waters. Settled just outside the Greensboro area in Randolph County, Liberty seemed like a quaint, peaceful town at first glance. But as I quickly learned, appearances can be deceiving, and the calm surface often hides a churning undercurrent of small-town politics and personal vendettas.

Looking back, I realize I made a critical mistake right from the start: I didn't do my due diligence. Unlike my previous position in Norwood, where I had thoroughly researched the community and its power dynamics, I rushed into the Liberty job without scratching beneath the surface. It's a lesson I hope you can learn from without having to experience it firsthand.

As I settled into my new role, I began to uncover a web of issues plaguing me long before my arrival. One of the most glaring problems centered around an officer designated as the department's drug investigator by the previous administration. Keeping a problematic officer focused on known criminals rather than harassing the general public seemed like a smart move. In reality, it was like giving matches to a pyromaniac and telling them only to burn down abandoned buildings.

This officer, a preacher in a small local congregation, had been given free rein to wear plain clothes, drive an unmarked car, and operate with minimal supervision. The result? He was still stopping motorists for minor infractions and abusing his position, all while hiding behind the veneer of undercover work.

To make matters worse, this officer had secured every Wednesday and Sunday off, ostensibly for his preaching duties. This arrangement was causing significant tension within the department, as other officers could never take those days off. It was a clear case of favoritism that needed to be addressed.

Chief Taylor Desk

One of my first acts as chief was implementing a new policy requiring me to approve all off-duty employment. This policy specifically prohibited certain types of work that could conflict with police duties, such as bail bondsman, pawn shop worker, or private investigator. The policy also stipulated that an officer's primary duty was to the police department, and they had to be able to leave their part-time job if called in for police work.

He was less than thrilled when I confronted our "preacher-officer" about his undisclosed part-time ministry work. But this confrontation was just the tip of the iceberg.

A few days later, a citizen came to me with a complaint that would expose the depth of this officer's misconduct. The woman explained that she had been pulled over the night before, and her car had been towed. The reason? The officer claimed he stopped her because her license plate was about to expire on the 15th of the month, and he "just wanted to let her know."

For those unfamiliar with traffic law, this is a clear violation of the Fourth Amendment. Police officers can't simply stop someone without reasonable suspicion or probable cause that a crime is being committed. An upcoming license plate expiration doesn't even come close to meeting that standard.

As the woman continued her story, it became clear that this stop was just the beginning of her ordeal. After pulling her over, the officer asked for her expired insurance card. He then ran her information through the DMV (Division of Motor Vehicle) database, which showed her operating without an active insurance policy. Based on this, he wrote her a ticket and had her car towed from the grocery store parking lot where she had pulled in.

Clearly of limited financial means, the woman was now stuck without transportation and facing mounting tow and storage fees she couldn't afford. My blood boiled as I listened to her story. This wasn't just a case of an officer overstepping his bounds; it was a gross abuse of power that directly harmed the citizens we were sworn to protect.

I assured the woman I would look into the matter and do what I could to help her. Then, I called the officer into my office for a conversation that would be the first of many confrontations.

When I asked why he had stopped the woman, he repeated the line about her license plate expiring soon. I challenged him to show me where in the North Carolina Motor Vehicle Law, Chapter 20, it stated that this was an appropriate reason for a traffic stop. Of course, he couldn't because no such provision exists.

The officer began to argue, insisting he had the right to make such stops. I shut him down quickly, reminding him that I teach Motor Vehicle Law at the police academy and that his actions were clearly outside the bounds of legal police conduct. Not only did he lack reasonable suspicion for the initial stop, but he also couldn't be certain about the insurance situation. DMV records are often outdated, and North Carolina doesn't require drivers to carry physical insurance cards.

I informed him that the department would pay for the woman's tow bill and storage fees and that he needed to void the citation he had written. Then, I took the opportunity to address the issue of his undisclosed part-time ministry work.

He bristled when I mentioned that he was living in a parsonage and employed as a minister without seeking the required permission. I clarified that while I didn't have a problem with his role as a preacher, his automatic claim to every Wednesday and Sunday off was unfair to his fellow officers. Going forward, I explained, he would only be able to take those days if no other officers requested them for vacation or training.

This conversation did not go over well, to put it mildly. The officer and his supporters, including his mother, who was running for town commissioner, began spreading misinformation about my policies. This incident sparked a political firestorm in the small town. They claimed I was against him being a preacher and was trying to prevent him from practicing his faith. It was a masterclass in small-town politics and rumor-mongering designed to turn public opinion against me.

When I thought the situation couldn't get more complicated, it took an unexpected and disturbing turn. About a week after my confrontation with the officer, my administrative assistant worriedly called me to her desk.

"Chief, I need to show you something," she said, her voice tight with concern. "I did not go to this website. I did not put this on my computer, but this is what happened."

She turned on her monitor, and I found myself staring at a homosexual pornographic website displaying explicit images of men engaged in sexual acts. It was the last thing I expected to see on a government computer in a small-town police department.

A quick check of the login history revealed that our problem officer—the preacher—had accessed the site using his credentials, and during his work hours, no less.

He denied everything when I confronted him about violating our IT policies and creating a hostile work environment. He claimed he had given his password to other people, a weak excuse that only compounded his misconduct.

At this point, I had no choice but to begin disciplinary proceedings. I recommended termination based on the improper towing incident, his failure to disclose his part-time job, and the pornography incident. The officer, predictably, lawyered up.

In a meeting with his attorney, they proposed that he stay on as a reserve officer, maintaining his rank and continuing to do whatever he wanted. I flatly refused. The best I was willing to offer was to let him resign, provide a neutral reference to future employers, and not move to revoke his law enforcement certification. They agreed with these options, and the officer left the department.

I thought that might be the end, but small towns have long memories and even longer grapevines. A few weeks later, three men from the community came to my office to apologize. They identified themselves as members of the former officer's congregation and expressed remorse for what they said about me based on his claims.

They described a recent incident at their church that had opened their eyes to the officer's true nature. Apparently, during a sermon, one of the men had been shaking his head in disagreement with something the former officer was saying. The ex-cop stopped his sermon, singled out the man, and threatened to have him arrested and dragged out of the church if he didn't stop.

This outburst shocked the congregation and led to the board of elders removing him as their pastor. While it was gratifying to have some vindication, the damage to my reputation in the community had already been done.

As the next round of elections for town commissioners approached, it became clear that my tenure in Liberty would be short-lived. The two allies I had on the town board, a local dentist and a businessman, decided not to run again, tired of dealing with the political turmoil. Their likely replacements included a man whose son we had arrested for trafficking marijuana and the mother of our problem ex-officer – both with clear vendettas against the police department.

Recognizing the writing on the wall, I began seeking other employment opportunities. It was a disappointing end to my time in Liberty. Still, it reinforced a crucial lesson: always do your due diligence before taking a position, especially in the politically charged world of small-town law enforcement.

This experience highlighted the typical lifespan of a police chief, which averages between three to five years. While exceptions exist, this turnover is often tied to the rotation of city council members or town commissioners. As new faces come into power with different ideas and focus areas, it frequently spells trouble for the existing police leadership.

Despite the challenges and disappointments, I still admire those who volunteer for public service in their communities. Whether serving on a town board, a homeowners association, or in local government, these roles subject individuals to intense scrutiny and criticism. I hope most are there for the right reasons – to foster a better environment for everyone.

As I left Liberty, I reflected on the complexities of balancing justice, community service, and small-town politics. It's a delicate act that requires integrity, adaptability, and a thick skin. While my time there was shorter than I had hoped, it provided valuable lessons that would serve me well in my future endeavors.

Shortly after leaving Liberty, I moved to apply my law enforcement experience to a broader public safety mission. This move took me to the Governor's Highway Safety Program, where I served as an external affairs manager overseeing several crucial safety initiatives: the "Click It or Ticket" program focused on occupant restraint, the "Booze It and Lose It" campaign targeting impaired driving, and the Safe Communities program aimed at accident prevention.

This role allowed me to continue interacting with law enforcement agencies, helping them secure grants for overtime enforcement and equipment. I traveled extensively across North Carolina, working to make our streets safer. One of my key strategies was to analyze data on the state's most deadly counties, identify patterns in accident locations, and then assemble multidisciplinary teams to address these issues.

These teams included local Department of Transportation engineers, emergency room doctors and nurses, and health and human services representatives. Together, we would assess whether the high accident rates were due to engineering problems, education gaps, or enforcement issues. This collaborative approach led to significant reductions in fatality crashes across the state.

One of our most successful initiatives was the implementation of the "Batmobiles" - mobile breath-analyzing testing stations. These converted buses, equipped with rooms for sobriety tests, breath tests, and even a magistrate, allowed for efficient processing of impaired drivers without overburdening local facilities.

While I found this work rewarding, after a year, I realized I missed being directly involved in community policing. This led me to my next position as the chief of the Dorothea Dix Hospital Police Department, one of three state psychiatric facilities in North Carolina.

This unique environment presented its own set of challenges. We had to balance the needs of patient care with public safety, often dealing with individuals who had severe psychiatric issues or were deemed criminally insane. We faced everything from patient elopements (unauthorized departures) to assaults on staff. We even had to coordinate with the nearby state prison for patients who became too dangerous for the hospital setting.

Throughout these experiences, from small-town politics to state-level initiatives and specialized institutional policing, I learned valuable lessons about leadership, public service, and the complexities of law enforcement. I realized that being an effective leader often means navigating choppy waters, balancing competing interests, and always striving to do what's right, even when it's unpopular.

Chapter 13

Emergency Management

There's a moment in every life given to service when you realize that your calling doesn't stop at the boundaries of a particular job or sector. The work may change, and the environment may shift, but the essence of serving others—being prepared and responsive—remains.

That moment of awareness arrived on a quiet Sunday afternoon in January 2000 as I casually scanned the help-wanted newspaper ads—a habit I'd developed even though I was comfortably employed. It was an unassuming moment, yet it became a turning point in hindsight. It was the call for the next chapter of my life of service.

At that time, I had spent years in public service, working within the framework of government roles. The idea of shifting to the private sector wasn't something I had considered seriously before. Yet, service is unpredictable, and when it calls, you answer.

As I scanned, I caught an intriguing advertisement from an unknown company with only a P.O. box, seeking "emergency response professionals." No company name, no detailed job description. Just those three words and a mailing address. The vagueness of the term piqued my curiosity. After all, with my diverse background as a police officer, security officer, firefighter, emergency medical technician, and even a bomb dog handler in the Air Force, what else could emergency response entail? I never knew it wasn't just about responding to emergencies; it was about leading, preparing, and communicating in ways that would keep thousands of people safe.

Driven by curiosity and ambition, I sent my resume to the mysterious address. Weeks passed, and I had almost forgotten about the application when I received a call from a hiring manager. What followed was a series of interviews that would challenge my expertise and ultimately lead me to a pivotal role in my career.

Preparing for RNC

The interview process was intense, to say the least. I faced not one but five different vice presidents at Nortel Networks. As I would soon learn, Nortel was a telecommunications giant riding high on the wave of the tech boom. They pioneered fiber optics, cleverly leveraging contracts with U.S. railways to lay cables along existing tracks and repurpose railroad infrastructure for their network components. It was a brilliant strategy, and in early 2000, Nortel stock seemed poised for another split, promising to make its investors even wealthier.

On the day of the interviews, I donned my best suit and tie, determined to make a strong first impression. As I navigated the sleek, modern offices of Nortel, I couldn't help but feel like I was stepping into the future. The atmosphere was electric, quite different from the often staid government buildings I was accustomed to.

I'll always remember when I sat down with one of the VPs. He looked me up and down, a smirk playing at the corners of his mouth, and said, "You know, this will be the first and last time you'll wear a tie to work here." It was a small comment, but it spoke volumes about the culture I would enter - less formal and more focused on results than appearances.

The interviews went well - surprisingly so. Despite the vast differences between my public sector background and this corporate environment, I could confidently answer their questions. My diverse experience in various emergency response roles impressed them, and before I knew it, I was asked to travel to Pennsylvania for another round of interviews.

In a nondescript conference room far from home, I truly began to understand the scope of the role they were considering me for. They peppered me with questions about emergency protocols, risk assessment, and crisis management. With each answer, I could see their eyes light up as if they were discovering exactly what they'd been looking for.

Weeks later, the offer came: I was hired as the East US Emergency Manager for Nortel Networks. My responsibility would span from New York to Florida and the East Coast to the Mississippi River. I would be tasked with developing and maintaining emergency response plans for all Nortel facilities in this vast territory.

I quickly realized that while the fundamental principles of emergency management remained the same, the context was entirely different. In the public sector, resources were often limited, and bureaucracy could slow response times. In the fast-paced world of a tech giant, efficiency was paramount, and resources were plentiful - if you could justify their need.

My role on paper wasn't unfamiliar, but it required me to approach things differently in practice. What struck me most during my time there was how vital communication is to individuals given to a life of service—and not just in the literal sense of talking to people but in ensuring that what we say and how we say it truly serves those we are responsible for.

My first major task was to create a comprehensive emergency action plan. This wasn't just any document; it would be the blueprint for how Nortel would respond to any conceivable emergency, from natural disasters to man-made crises. I threw myself into the work, drawing on every ounce of experience I'd gained.

The plan I developed was a masterpiece of preparedness. It covered everything from medical emergencies to hazardous material spills, severe weather protocols, and bomb threats. But I quickly realized something critical: It didn't matter how perfect the plan was if people couldn't understand or access it when needed. So, I created a quick reference guide that sat by every phone in our facilities. The information was distilled into the essentials—emergency numbers, immediate steps, and clear, simple language. In an emergency, clarity saves lives.

This idea of clear communication wasn't new to me. In law enforcement, we used "10 codes" on the radio to relay information quickly. However, these codes could sometimes cause confusion, especially when working with multiple agencies using different codes. What might mean "officer needs assistance" in one jurisdiction could mean "lunch break" in another. In emergencies, this small misunderstanding can lead to disaster. At its core, service demands that we communicate in ways that reduce confusion and empower others to act confidently.

This lesson was driven home during my time at Nortel, especially when we faced a series of bomb threats at the Research Triangle Park facility. Before I was hired, the Gateway facility—a sprawling complex of over 1 million square feet—had received no less than eight bomb threats. My job was to devise a plan to deal with these threats and identify the person making them.

The threats kept coming, and each time, they disrupted our operations. At first, evacuations seemed the best response, but something didn't sit right with me. I approached the problem like a detective, analyzing the pattern of the calls. They all came just before lunch on Fridays, originating from a section of the building undergoing renovation.

This told me two things: first, that the caller was likely looking to get out of work early on Fridays for personal reasons, and second, that they had access to a relatively isolated area of the building.

The area in question was stripped bare except for the telephones. It was covered in construction debris and dust, making gathering physical evidence, like fingerprints, nearly impossible. The calls came from the same general area but not the same phone, suggesting the caller was moving around to avoid detection.

After careful consideration, I boldly recommended to management that no more evacuations be made due to bomb threats. Instead, I proposed organizing search teams in all areas near the impacted zone. My reasoning was based on my experience with explosives - to do significant damage to a building of that size, you'd need a massive bomb, something that would be hard to conceal.

Moreover, I was concerned about the designated rally points where people gathered after an evacuation. In the event of a real threat, these gathering spots could become prime targets for a vehicle-borne improvised explosive device. It was a sobering thought but one that needed to be considered. I had to be clear with management and communicate what we were doing and why it needed to change.

We shifted our response, conducting careful searches instead of mass evacuations, ensuring our actions truly protected people. In the end, the bomb threats stopped, and no one was hurt. It was a reminder that in service, leadership means looking beyond the obvious solution. It means communicating your reasoning clearly, even when it challenges the status quo, because the safety and well-being of those you serve must always come first.

One of the most vivid experiences from this period was a full-scale emergency drill we conducted. Our fire evacuation and medical response teams were tested in real-time with a staged gas leak and "victims" needing rescue. Watching our team's initial response, I saw mistakes that could have been catastrophic in a real incident. They rushed in without adequately assessing the scene, putting themselves in danger. Only after repeated failures did one team recognize the environmental hazard and call in the Hazardous Material Unit (HAZMAT). It was a humbling moment for everyone involved.

The drill wasn't just about running a simulation—it was about communicating the importance of being thorough and thinking clearly in the heat of the moment. Had they taken a moment to assess the situation, they would have noticed the warning signs about inhalation danger. We held debriefs afterward, and while the mistakes may have disheartened some, I saw it as a valuable opportunity. See, service is about constantly learning and improving, and communication is at the heart of that process. It's not just about reacting at the moment but about preparing others to be better prepared themselves.

One incident during the after-action meetings sticks with me to this day. I sent a follow-up email to the head nurse, asking her team's feedback on the drill. Her response was just two letters, "FU," which stunned me. Our working relationship had been positive, so this seemed out of character. I went to her office to talk in person, and after a few minutes of confusion, she burst into laughter. "I meant 'follow up'!" she explained. It was a simple misunderstanding, but it underscored a truth: We must never take communication for granted in service. What we say might be clear to us, but we must ensure it's clear to others.

That incident wasn't the last time I'd find myself caught in the crossfire of confusing acronyms. Allow me to share another enlightening - and somewhat embarrassing - experience that further drove home the importance of clear communication in any field, especially one as critical as emergency management.

As the new emergency response manager at Nortel Networks, one of my duties was to work closely with the nurses, ensuring they were prepared to respond to medical emergencies in our facilities. Their primary responsibilities included overseeing employee health and welfare, clearing international travel medically, and keeping immunizations current.

I vividly remember the first meeting I was invited to. I was eager to introduce myself and participate in the nurse's information-sharing sessions. As I sat there, trying to absorb all the new information, I heard one of the presenters mention that there had been an increase in STDs. My ears perked up immediately. "STDs?" I thought, "That's a rather personal topic for a workplace meeting."

Confused but not wanting to embarrass myself as the newcomer, I kept my mouth shut and continued listening, filing this unusual tidbit of information away for later consideration. However, when the same presenter mentioned an increase in STDs at the following month's meeting, I couldn't contain my curiosity any longer.

Feeling more comfortable in my role and making new friends among the staff, I blurted out, "What are these people doing? Are they having orgies or what? How are we getting these increases in STDs?"

The presenter looked at me, bewildered. "What are you talking about?" she asked.

"How are we getting all these sexually transmitted diseases?" I clarified, feeling increasingly uncomfortable as all eyes in the room turned to me.

There was a moment of stunned silence before the presenter burst into laughter. "No, no," she managed between chuckles, "STD stands for short-term disabilities!"

As the realization dawned on me, I felt my face flush red with embarrassment. The room erupted in good-natured laughter, and I couldn't help but join in despite my mortification. It was a humbling moment, but it was another reminder that in service—especially in leadership—plain, clear language is always the best path.

As I settled into my role at Nortel Networks, I contemplated my future career trajectory. The private sector was a new world for me, with its unique structure and expectations. It didn't take long for me to realize that if I wanted to climb the corporate ladder, I needed to broaden my skill set.

In the corporate world, I discovered that there typically isn't a security or safety vice president. Instead, these roles are often combined into a single position: vice president of safety and security. The reason for this, I learned, was rooted in the bottom line. Neither security nor safety are profit centers for an organization; they don't bring money in. Instead, their value lies in reducing money going out, preventing theft and loss, or minimizing workplace injuries and the associated costs.

This realization led me to an important question: How could I better prepare myself for my next move after Nortel? While I had years of experience in law enforcement and security-related functions, I recognized a gap in my knowledge regarding occupational safety.

After some research, I discovered that East Carolina University in Greenville, North Carolina, offered the only accredited occupational safety master's program in the country at the time. It seemed like the perfect opportunity to round out my skill set. With my bachelor's degree in criminal justice and my technical certifications in law enforcement and security, I believe that adding a master's in occupational safety would position me well for future advancement in corporate America.

Based on these findings, I enrolled in the program, balancing my studies with my work at Nortel. Many of the classes were virtual, which allowed me to maintain my full-time job while pursuing my degree. The curriculum was eye-opening, covering various topics, including OSHA regulations and their impact on the work environment, ergonomics, hazardous materials management, and more.

Juggling work responsibilities, family life, and graduate studies was challenging, but I was driven by the knowledge that this education would open new doors for me in the future. In December 2003, my hard work paid off when I graduated from the program with honors. Adding a "Master's Degree in Occupational Safety" to my resume felt like a significant milestone that would set me apart in future job searches.

This experience reinforced a piece of advice I now share with anyone looking to advance their career: research the requirements for the type of job you're seeking, not just at the entry level but at the executive level. Understanding what qualifications are needed at the top can help you chart a course for your professional development.

I'm a firm believer in the value of formal education. I often refer to the "McDonald's application" - that initial application all entities give you to fill out. It typically asks about your education, with checkboxes for high school diploma, GED, associate's degree, bachelor's degree, master's degree, and doctorate. The more boxes you can check, your application will likely move from the HR department to the hiring manager's desk.

But formal education is just the beginning. Technical education and specific skills are equally important. Many applications ask about skills you've acquired, whether it's typing speed, language proficiency, computer programs you're familiar with, or, in the case of law enforcement, different types of investigative techniques or specialized tools.

Lastly, it's crucial to understand how your previous jobs relate to where you're trying to go. Even if a past position isn't directly in line with your goal, there's almost always some component of your former job description that you can relate to your target role. The key is to tailor each application to the position you're aiming for by highlighting the most relevant experiences and skills.

Looking back, my time at Nortel was about more than managing emergencies. It was about learning to serve people by ensuring they had the right tools, information, and confidence to act when it mattered most. Service is not just about responding to crises—it's about preparing yourself and others. It's about communicating clearly and ensuring everyone feels empowered to make the right decisions.

As I continued my career, I carried these lessons with me. Whether working with the local fire departments to update our hazardous materials protocols or supporting the Massachusetts State Police in their bomb disposal training, my focus was always on one thing: How can I serve in a way that makes others stronger? How can I communicate so that those I'm responsible for have everything they need to be safe and effective?

Service, at its core, is about putting others first. It's about being transparent, consistent, and always willing to go the extra mile to ensure the people around you understand what to do and why it matters. That's the heart of leadership—whether in the public sector, private industry, or any walk of life.

I've often found that the most influential acts of service are the ones that happen quietly, behind the scenes. The emergency plan you write that might never need to be used. The conversations you have that clarify misunderstandings before they become problems. The drills you conduct reveal weaknesses so they can be strengthened. These are the moments where service truly shines.

And in all of it, our most essential tool is communication. Clear, honest, and direct communication can turn plans into action, confusion into clarity, and ordinary people into capable leaders. That's what service is all about—not just doing the work ourselves but ensuring that others are ready to take up the call when it's their turn.

This is the heart of a life of service. We don't serve for recognition or reward. We serve because it's who we are. In that service, we strive to leave every place, person, and situation better than we found. This commitment to clear communication, preparedness, and the well-being of others defines us.

Chapter 14

Iraq: Call to Duty

After leaving my position as a full-time police chief, I still had a burning desire to serve my community, but I wasn't sure how to channel that energy. Although I was still working as an emergency coordinator at Nortel, I started considering the National Guard as an additional place to serve. The Guard was about responding to local emergencies - helping out during winter storms, hurricanes, or other disasters. I saw it as more of a humanitarian organization, a way to be there for my neighbors when they needed it most.

I had little idea that my journey with the National Guard would take me far beyond the borders of North Carolina and into the heart of a war zone. But let's not get ahead of ourselves. Every great adventure and service starts with a single step, and mine began with a visit to a local recruiter's office.

When I walked in, full of confidence from my 30 years in law enforcement, I told the young recruiter that I wasn't interested in becoming a military police officer. I've been there and done that, you know? I was looking for something different, maybe something in aviation. I've always loved to fly, though at 39, I knew becoming a pilot was out of the question. There had to be other interesting specialties, right?

The recruiter, bless his heart, looked at me like I'd just spoken Greek. He told me we couldn't even have an honest conversation until I took the ASVAB - the Armed Services Vocational Aptitude Battery test. I chuckled and told him I'd taken that test in high school and qualified for every job in the military except for some in finance. But rules are rules, so I agreed to retake it.

After acing the test (if I say so myself), I thought we'd finally get down to business. But no, there was another hurdle - the physical. I explained that I was part of a hazmat team and took an extensive physical every year, far more rigorous than what the army would require. But again, rules are rules. So, I submitted to the exam, passed with flying colors, and finally sat down with the recruiter, ready to discuss my future as an officer in the National Guard.

That's when things took a turn. The recruiter started talking about openings in transportation, logistics, and ordnance. Now, for those who don't speak Army, ordnance in this context means mechanical work, like fixing trucks. I'll be honest, I was disappointed. This wasn't what I had in mind at all. We could have had that conversation early on and I would have been on my merry way. I told the recruiter he'd wasted my time and left feeling slightly deflated.

But fate had other plans. The next day, I got a call from the NCOIC—the Non-Commissioned Officer in Charge, the head honcho of the recruiters. He'd looked at my scores and background and thought I might be interested in counterintelligence.

Now, that caught my attention. I knew the military had intelligence units, but counterintelligence? That was new to me. The NCOIC sent over a job description, and let me tell you, it was like Christmas morning. Counterintelligence agents investigate sedition, treason, and espionage for the army. They often work in plain clothes or wear uniforms without rank. They carry guns and have law enforcement credentials. All my years in law enforcement, I had been leading up to this moment.

I was sold, but the NCOIC had one more suggestion. Given my age and experience, he thought I'd be better off going in as a warrant officer.

Warrant officers, he explained, do most of the fun work - surveillance, investigations, the real nitty-gritty stuff. Officers, on the other hand, often get stuck with administrative duties. It sounded perfect.

So, I signed the contract. The plan was for me to go in initially as an E4 specialist (that's enlisted rank), attend warrant officer school, get commissioned, and finally head off to counterintelligence school at Fort Huachuca, Arizona.

As I resumed my duty station, my first drill with the 139th ROC (Rear Operations Center) in Morrisville, North Carolina, was eye-opening. I showed up in my E4 uniform, expecting a secure facility with guards and locked doors. Instead, I wandered around an unsecured building, trying to find someone who could point me in the right direction. It was a far cry from the high-security world I'd imagined, but hey, you've got to start somewhere, right?

The real excitement began the following month as I packed my stuff and was transported to Fort Bragg for our annual training. We set up shop at the John F. Kennedy Warfighter School, creating a tactical operations center complete with maps and intel for a mock battle we would be involved with for the next 15 days. The operation was centered around Tunisia. It was like a giant, high-stakes game of Risk, and I loved every minute.

As we set up our board one afternoon, I explained how I thought the enemy would attack my fellow soldiers. I pointed out their likely movements, where they'd need to resupply, and where I expected them to hit us. What I didn't realize was that the colonel in charge of the ROC was sitting right behind me, listening to every word.

Over the next week, the mock battle unfolded precisely as I'd predicted. At the end of the exercise, the colonel called me into his office. "How did you know what was going to happen?" he asked. "Do you have somebody in the white cell?" Now, I had no idea what a white cell was - remember, this was only my first drill. I explained that I was applying the same logic we use in law enforcement, looking at the enemy's movements the way we'd analyze criminal activity.

The colonel was impressed—so impressed, in fact, that he decided I wasn't going to be a warrant officer after all. "You're going to be an officer," he told me. We're getting a waiver, and you're going to officer candidate school." I was stunned. It was an unexpected honor.

And just like that, my path changed again. Over the next 12 months, I prepared for Officer Candidate School (OCS) at Fort Bragg. In March 2001, I started pre-OCS, a program designed to give prospective candidates a taste of what would come and help with physical conditioning.

I'll never forget that cold March morning on the PT (physical training) field. Some younger candidates tried to take charge, insisting we all dress the same. I was prepared for the weather with my military-issued toboggan cap, winter PT gear, outer coat, and gloves. But some of the others were in T-shirts and shorts. Despite my protests, they relented, and we all ended up stripping down to our summer PT uniforms. Let me tell you, it was freezing!

As we lined up for our two-mile run, one of the instructors asked who thought they could finish in 12 minutes. Nobody raised their hand. Then he asked about 13 minutes, and I raised mine. The instructor looked at me skeptically. "Grandpa, you're going to run it in 13 minutes?" he asked. "Yes, Sergeant," I replied.

Well, wouldn't you know it? I finished in 13 minutes and seconds and was the first across the finish line. Not bad for the oldest candidate in the class, huh? My new friend Darren Webb, about eight years my junior, wasn't far behind.

July 2001 marked the official start of our 12-month OCS course. We kicked things off with annual training at Fort Jackson, South Carolina. OCS Class 44 started with 44 candidates, all anxious and excited as we loaded onto the bus. We'd heard stories about what to expect, including getting "smoked" (that's military speak for intense physical training) at the state line by our TACs (our instructors).

Fort Jackson was intense, to put it mildly. For most of the 15 days, we lived in the woods, setting up perimeters and sheltering in half-tents like a hostile environment. We attended classes during the day and worked late into the night. Sleep deprivation was part of the training, pushing us to our limits physically and mentally.

One night, severe weather rolled in with lightning strikes and possible tornadoes. We were moved to a new barracks that had just been opened for the South Carolina National Guard. The storm came through and did a lot of damage to our bivouac area, scattering our gear. The next day, we were bussed back out to it and began recovering all our gear scattered over quite a distance of the area.

We brought the equipment back to clean all the mud off it, only to discover there were no outside hose bibs on the new building. As we washed them in the shower and laundry room, we clogged the indoor drains with pine straw and debris, which, let me tell you, did not make us popular with the higher-ups. We were in more trouble for doing that than anything else.

At the end of those 15 days, we were all glad to get back on the bus to North Carolina. And yes, just as the legends foretold, we did stop at the state line on the way back. Our TACs met us there and had us doing push-ups, sit-ups, and flutter kicks in front of everyone at the North Carolina Welcome Center. Talk about a warm welcome home!

The following 11 months were a whirlwind of monthly weekend drills. We'd arrive at the barracks on Friday, set up our display racks (everything had to be displayed precisely and in perfect condition for Saturday morning inspections), then dive into a packed schedule of physical fitness, inspections, classroom work, and field maneuvers.

Then came September 11, 2001. The day that changed everything. Our nation was attacked, and many Americans were killed. The military was on high alert, and everybody was in a state of high readiness.

As class president, my phone rang with worried candidates asking if we were going to war. I did my best to calm their nerves, explaining that even if we were federalized and rushed through our training, it would still be at least six to eight months before we could be deployed overseas.

Despite the national crisis, our training continued as scheduled. We spent 15 days at Fort Benning, Georgia, running infantry lanes and practicing Army tactics. When we returned to Fort Bragg for our graduation, only 13 of the original 44 candidates received their commission that day. I was fortunate to be one of the honor graduates, and as class president, I was asked to give the final speech.

Now, here's where things got interesting. The TACs wanted to approve my speech in advance, but they didn't like it when I submitted my original draft. They ordered me to rewrite it according to their specifications. I complied... sort of. I wrote a new speech to their liking, knowing I wouldn't give it.

When graduation day came, I pulled out my original speech and began with the story of an Army Lieutenant Colonel who had been shot during PT at Fort Bragg in October 1995. Lt. Col. Guy LoFaro was shot by a soldier with neo-Nazi sympathies. The colonel's story, which he had shared with West Point students, used the Army values to explain the impact on his career. It was a powerful message about loyalty, pride, and perseverance in adversity.

My speech concluded with the idea that while our careers might not be as eventful as the colonel's, we would all strive to live up to the same Army values and take care of our soldiers. To my relief and pride, the speech received a standing ovation. Afterward, three general officers presented me with one of their challenge coins - a significant honor in military culture.

However, my military career was about to take an unexpected turn. While at Fort Benning, three full-bird colonels informed me that I wouldn't be branching into counterintelligence after all. Instead, I had to choose between ordnance, quartermaster, or transportation.

Ironically, these were the very specialties I had initially rejected. It was a bitter pill to swallow, but that's the Army for you - the needs of the service come first. I chose transportation and was assigned to the Corps Support Group in Durham, North Carolina. It wasn't the counterintelligence role I had initially dreamed of, but I was determined to make the most of it.

Shortly after my assignment, the unit received mobilization orders to support a Special Forces team in Afghanistan. We had just 90 days to prepare for deployment. Despite not completing my officer basic course, I was placed on active duty to assist with the mobilization preparations. The military often does this to ensure all hands are on deck during critical periods, even if some personnel haven't completed all their training.

Soon afterward, I was tasked with the position of public information officer (PIO). In this capacity, I found myself holding news conferences with local media. This wasn't just about informing the public; it was a strategic move by the military. You see, the National Guard is deeply rooted in local communities. Our soldiers are your neighbors, coworkers, and friends. We maintained that crucial connection between the military and civilian worlds by informing the public about our mission, timeline, and preparations.

I remember my first press conference. Standing before the cameras, I felt the weight of responsibility on my shoulders. "Our unit has been called to support our nation's efforts in Afghanistan," I announced, trying to keep my voice steady. "We'll be deploying within 90 days, and our primary mission will be to provide logistical support to Special Forces operations."

A reporter raised her hand. "How long will the deployment last?" she asked.

"At this time, we're looking at a 12-month deployment," I replied, "but as you know, these things can change based on the needs of the mission."

These press conferences weren't just about disseminating information. They were about reassuring families, preparing the community for the absence of their citizen soldiers, and fostering a sense of pride and support for our mission. It was a crash course in public relations, yet I had prepared for it. My experience as a police chief, where I had developed the skills to address the public and protect confidential information, served me well during this period.

When the main unit finally departed for Afghanistan, I was left behind as the rear echelon commander. Now, let me tell you, this wasn't exactly a coveted position. I was essentially in charge of what we affectionately called the "broke, lame, and lazy" - soldiers who couldn't deploy due to physical or administrative reasons. These weren't typically the most motivated troops, but I saw it as an opportunity to make a real difference.

At our first monthly drill, I laid out my goals: "Listen up," I said, looking each of them in the eye. "I know you're here because, for various reasons, you couldn't deploy with the main unit. But that doesn't mean you're not valuable soldiers. Our goal is to get everyone on the weight control program within standards, ensure all annual physical fitness test failures will pass, and accomplish any other missions that come our way. It won't be easy, but I believe in every one of you. Are you with me?"

A chorus of "Hooah!" echoed through the room. It wasn't enthusiastic, but it was a start.

True to my word, I implemented a rigorous physical fitness program. We trained hard, with PT sessions on Saturday mornings and Sunday afternoons. It was tough, but slowly, we started to see results.

We were able to complete our annual firearms qualification with the M16 rifle. In the military, marksmanship isn't just a skill – it's a fundamental aspect of being a soldier. Regardless of their role, every service member must be proficient with their weapon. It's about readiness, about being prepared to defend yourself and your fellow soldiers if the situation demands it.

I secured ammunition and range time, first for practice and then for qualification. To my satisfaction, all the soldiers passed the rifle qualification. But I wasn't content to stop there. When I learned about an upcoming statewide marksmanship competition, I saw a chance to push my troops and boost their morale.

We dedicated every drill weekend to honing our marksmanship skills for the next two months. We practiced breathing techniques, trigger control, and sight alignment. We ran drills in various positions—standing, kneeling, and prone. The soldiers grumbled initially, but their confidence grew as their skills improved.

The day of the competition arrived, and I couldn't have been prouder of my team. We managed to secure third place out of all the non-deployed units in the state. Seeing those soldiers, who had been left behind and possibly feeling demoralized, now standing tall at the awards ceremony was a moment I'll never forget.

Six months later, the group commander was impressed with our improvements when the unit returned from deployment due to changing mission requirements. All our soldiers were now within height and weight standards, everyone was passing their annual physical fitness tests, and we had that third-place trophy in marksmanship to show for our efforts.

This proved what could be achieved with dedicated leadership and a committed team. I drew on every skill I'd learned throughout my life - from Boy Scouts and Civil Air Patrol to my time in the Air Force and law enforcement - to lead this group and turn it into a cohesive, capable unit.

When I thought things were settling down, I received orders to attend the Transportation Officer Basic Course at Fort Eustis, Virginia. There, I reconnected with my friend from OCS, Lieutenant Darren Webb. A week after completing our training, Darren called me to ask if I'd heard anything about deploying. I was confused - my unit had just returned from deployment, so I didn't expect we'd be going anywhere soon.

"No, I haven't heard anything," I told Darren. "Why do you ask?"

"Well," he replied, "I had Captain Herbert Judon, the commander of the 1450th Transportation Company, ask me about you. He wanted to know what kind of officer you were."

I was surprised. "What did you tell him?"

"I told him you were a fine officer and a great leader, that you'd be an asset to any unit."

I thanked Darren for the compliment, but my mind was racing. Why would another unit's commander be asking about me?

When I followed up with my command staff, I learned that there was indeed a "drug deal" (military slang for an unconventional arrangement) being made to have me transferred to a unit out west. In the Army, these informal arrangements are often how things get done, especially when there's an urgent need to fill a specific role.

Before I knew it, I was assigned to the 1450th Transportation Company in Lenoir, North Carolina, after receiving orders to mobilize to Iraq. It was December 2003, and my world was about to change dramatically as the sands of Iraq awaited, and with them, challenges and experiences that would shape the rest of my life and career.

Chapter 15

The Bronze Star

In January 2004, we were placed under federal orders to deploy to Iraq for line-haul operations. I was initially assigned to 1st Platoon, a typical configuration of about 43 soldiers led by a lieutenant (me), with a sergeant first class as the platoon sergeant, four staff sergeants as squad leaders, and further subdivisions led by sergeants.

However, the company commander soon approached me with a challenge. "Lieutenant," he said, pulling me aside one day, "I've got a situation with 3rd Platoon. The lieutenant I assigned there is having... difficulties."

"What kind of difficulties, sir?" I asked.

He sighed. "He feels the soldiers are too 'inbred' for him to manage effectively. They all come from the same small town and mostly work for the same telephone company. He can't seem to connect with them."

I nodded, beginning to understand. In the National Guard, it's not uncommon to have units where many soldiers come from the same community or workplace. This can create strong bonds but can also be challenging for an outsider to lead.

"I was wondering," the captain continued, "if you'd consider changing platoons. See if you can have a positive influence on the 3rd Platoon."

I didn't hesitate. "Of course, sir. I'd be happy to try."

It turned out to be one of the best decisions of my military career. Far from being a problem, the tight-knit nature of the 3rd Platoon was its greatest strength. These soldiers had lifelong friendships dating back to elementary school and high school and years of working together. They already had a strong organizational structure and took care of each other in a way that you couldn't ask for in a typical unit.

This preexisting cohesion made my task as their leader much easier. I didn't have to build esprit de corps or a sense of teamwork—it was already there. My job was to harness and direct that energy, and I was proud to lead this exceptional group of soldiers.

The following month was a whirlwind of activity as we prepared for deployment. We packed crates of unit equipment, meticulously categorizing and logging everything. Our equipment would be shipped by rail to a port, loaded onto a ship bound for Kuwait, and offloaded there.

Next came the task of preparing our vehicles for shipment. As a line haul transportation company, our primary vehicles were M-915s - the Army's equivalent of civilian semi-trucks, painted olive drab. These weren't new vehicles; they had seen action in Desert Storm and were well-used.

"Alright," I addressed my platoon one morning. "These trucks have been through a war already. It's our job to make sure they're ready for another one. I want every nut, bolt, and wire checked and double-checked. Our lives may depend on these vehicles, so let's ensure they're in top shape."

We performed thorough preventive maintenance, including checking everything from tire pressure to engine oil, brake systems, and electrical components. It was tedious work but essential. In a combat zone, a breakdown could be more than just an inconvenience—it could be life-threatening.

The process of loading our equipment was a logistical challenge in itself. We stacked our flatbed trailers, carefully backing one on top of another, a tedious task that my soldiers handled with impressive skill.

"Easy does it!" I called out as one of my soldiers maneuvered a trailer into position. "Remember, slow is smooth, and smooth is fast."

Everything was then secured to rail cars, and we watched as the train departed for the port, carrying the tools we would soon rely on in a war zone.

With our equipment on its way, we loaded up in our assigned Humvees, and the rest of the soldiers boarded buses. We formed a convoy and went to Fort Stewart, Georgia, for our pre-deployment training.

The living conditions at Fort Stewart were far from luxurious. We were housed in crude, one-story concrete barracks with open latrines. Privacy became a distant memory as we adjusted to communal living - a row of toilets without dividers, urinals and sinks lining the walls, and large group showers reminiscent of high school locker rooms.

Our days at Fort Stewart were filled with intensive training. We covered many topics: the rules and laws of war, operating our equipment in a combat environment, more rifle qualifications, and learning critical skills like checking for unexploded ordnance.

I'll never forget the surreal experience of practicing how to probe for mines using a plastic fork from our MREs (Meals Ready-to-Eat). "Remember," our instructor said, demonstrating the technique, "slow and steady. You're feeling for any resistance. Hit something solid, you stop immediately and call EOD (Explosive Ordnance Disposal)."

It was a sobering reminder of the dangers we would face. This wasn't just a training exercise – it was preparation for situations where a wrong move could cost lives.

By the end of February, our training was complete. The unit was signed off as proficient and ready for deployment. We returned to Fort Bragg (now called Fort Liberty) for a yellow ribbon ceremony with our families.

Saying goodbye to loved ones was bittersweet, as we did not know when or if we'd see them again.

After the ceremony, we boarded a plane for Kuwait. The flight took about 28 hours, briefly stopping at Heathrow Airport in England. We finally landed in Kuwait at midnight and were shuttled to Camp Arifjan, our temporary headquarters. The next few weeks were filled with additional training on local customs and courtesies, which is crucial information for operating in this new environment.

Finally, the ships carrying our equipment arrived. We spent an entire day at the port, driving our vehicles off the massive cargo ships and convoying them to our new motor pool at Camp Arifjan. It was exhausting work, but by 7:30 PM, we had our motor pool set up, with rows of vehicles parked in perfect military order – a slight touch of order in the deployment chaos.

After a long day of work, I finally hit my cot, exhausted but satisfied with our progress. It felt like I had barely closed my eyes when my platoon sergeant shook me awake. "Sir, wake up! We've got a mission!"

I blinked, trying to clear the fog from my mind. "What do you mean, we've got a mission?"

"We're pushing out 150 trucks in the morning," he replied. "We have to get them ready."

The news hit me like a bucket of cold water. The enormity of the task ahead was staggering. We had just arrived - we had no maps, ammunition, or way to communicate with local resources. Our equipment hadn't even been through a technical inspection yet.

This wasn't how things were supposed to work, but it was the reality we faced. Plans often go out the window in war, and you must adapt on the fly. One of the most challenging aspects of military leadership is being ready to act decisively even when you don't have all the necessary information or resources.

We worked through the night, scrounging equipment from other units, cobbling crude maps, and gathering whatever intel we could about what to expect. As dawn broke, we found ourselves escorting a convoy of foreign national drivers in their Mercedes trucks, tasked with guiding them safely through enemy territory to deliver crucial supplies and equipment.

Our navigation tool? A printout of a PowerPoint slide with a crude, hand-drawn map. It differed significantly from the gridded maps and precise coordinates I'd been trained to use. The directions were vague at best – "Drive 120 miles on this main supply route; at approximately this mileage, you'll see two large television antennas; take the next exit."

As we set out on our first mission, I couldn't help but think of Private Jessica Lynch, who had been captured during Operation Desert Storm after her vehicle made a wrong turn. With our crude directions, I realized how easily that could happen to us. Every intersection became a moment of doubt - were we on the right path or one wrong turn away from disaster?

To add to our challenges, we had no translators, making communication with our Pakistani drivers nearly impossible. These men, essentially indentured servants to Kuwaiti contractors, were as much in the dark as we were.

Despite these obstacles, we pushed on. Every mile covered was a small victory; every checkpoint passed a relief. The constant vigilance was exhausting but necessary. We knew that danger could come from any direction at any moment. Through skill, determination, and perhaps luck, we completed our first mission and returned to base safely with no injuries. It was a baptism by fire, but we had proven ourselves capable.

Little did we know that April 2004 would prove to be a brutal awakening to the realities of war, with over 140 U.S. service members losing their lives on Iraqi soil. It was the bloodiest month since the conflict began, with attacks on our convoys becoming a near-daily occurrence.

Our woefully inadequate protection hammered home the gravity of our situation. My Humvee, like most, was little more than a target on wheels—canvas doors, a fabric roof, and whatever improvised armor we could cobble together. I'd lined the floor with old sandbags and spare flak vests, a pitiful defense against the increasingly sophisticated IEDs our enemies were deploying.

As the tempo of operations increased, so did the complexity of our missions. I was tasked with a particularly challenging assignment—a two-pronged operation that would stretch our resources to the limit. The first part involved delivering materials to Logistics Base Anaconda in northern Iraq. The second, arguably more crucial, was transporting 102 brand-new up-armored Humvees to an infantry unit in the desert near the Syrian border.

After careful consideration, I decided to split the mission. My platoon sergeant, a seasoned veteran I trusted implicitly, would handle the Anaconda run. We'd been there before, and I was confident in his ability to execute that leg without issues. I, on the other hand, would lead the Humvee delivery personally.

Now, let me explain what a Humvee is for those who might not be familiar. It stands for High Mobility Multi-Wheel Vehicle—H for high, M for mobility, M for multi, W for wheel, and V for vehicle. We weren't just delivering any Humvees; they were the up-armored variants desperately needed by our infantry units facing the brunt of the insurgency.

Our journey began at Forward Operating Base Cedar II, where we rested for the night before setting out. I watched my platoon sergeant's convoy disappear into the distance, heading for Anaconda, before turning my attention to our departure. That's when I hit my first snag.

The transportation office, in its wisdom, refused to let us leave. Despite having five military vehicles, my Humvee, and 59 foreign national-driven trucks, we didn't have enough "gun trucks"—Humvees, five-tons, or bobtail 915s specifically tasked with convoy protection. It was a bureaucratic nightmare that made you want to pull your hair out in frustration.

Determined not to let red tape derail our mission, I started working on the problem. I approached other convoy commanders, trying to piggyback onto their operations for the required gun truck quota. Just when I was about to give up hope, I met a character straight out of a war movie—a slightly unhinged Army Reserve lieutenant colonel who agreed to let us join his larger unit. He had an overabundance of gun trucks.

The next morning, we finally got clearance to roll out. Once outside the gates, we split off from the colonel's convoy and set our sights on our destination - Forward Operating Base Duke.

FOB Duke was a clear reminder of the harsh realities of frontline deployments. Apart from a single captured Iraqi Army building, the entire base was a sea of tents - shelter halves paired up to create cramped living quarters for the troops. A mobile AFES (Armed Forces Exchange Service) trailer and a mess and shower tent served as the base's convenience store.

Upon arrival, I went to the Tactical Operations Center and introduced myself to the colonel in charge. "Sir," I said, "I've got 102 new up-armored Humvees for you. The only problem is I don't have a way to offload them. Our trucks don't have ramps—they're standard flatbeds."

The colonel's response was quintessentially military: "Carry on, Lieutenant. Do what you need to do."

What followed was a master class in battlefield improvisation. We managed to get our hands on a backhoe and used it to build up the berm surrounding the helicopter landing pad. It was like playing with the world's largest, most frustrating sandbox. The sand kept collapsing, forcing us to pile more on top continuously.

The next scene was something out of an old episode of "Rat Patrol." We jumped these brand-new Humvees off the trailers onto our makeshift sand ramps. I could see the barely concealed grins on my soldiers' faces - for a moment, the danger and stress of our situation were forgotten in the sheer absurdity of what we were doing.

With all vehicles finally offloaded and secured, I reported to the commander that the mission had been accomplished. I was hoping for some guidance back to the MSR (main supply route) so we could head home. Instead, I got a curveball.

The colonel fixed me with a look that spelled trouble. "Lieutenant," he said, "we're about to take out some weapons caches. I could use you and your soldiers on these raids."

I straightened up, ready to volunteer. "Glad to help, sir. But it's just me, my driver, my gunner, and 59 foreign nationals."

The colonel's expression shifted from excitement to confusion to barely concealed exasperation. "Lieutenant, keep your foreign nationals outside my gate. I'll get you out of here when I can."

And just like that, I found myself in a surreal situation. For the next 20 days, I lived in a 40-foot conex - a metal shipping container - in the middle of the desert, along with 59 Pakistani drivers. Despite the language barrier (we had no translators, and their English was minimal at best), we managed to form a sort of camaraderie.

Our entertainment consisted of watching movies on my portable DVD player, its small screen a window into a world far removed from our dusty, dangerous reality. Our sustenance came from old cases of water and MREs mysteriously stacked outside the base - not exactly five-star dining, but it kept us going.

I'd enter the Tactical Operations Center for the Commander's Update Brief (CUB) every morning like clockwork. And every morning, I'd ask the colonel if he could spare a couple of soldiers to guide us back to the main supply route. Every morning, the answer was the same: "No."

Finally, on April 20th, 2004, I heard the words I'd been waiting for. "Lieutenant, I'm going to get you out of here today," the colonel announced. He assigned two NCOs to lead us back to the MSR.

The sergeants briefed me on their planned route, which included crossing a river via a bridge. My instincts kicked in. "Have you personally inspected this bridge?" I asked. "Are you sure it can support the weight of our M915 trucks and trailers?"

They assured me the bridge was up to the task. I should have known better than to take their word for it.

I felt relief and apprehension as we formed our convoy and began our journey home. Little did I know, this was just the beginning of what would become one of the most harrowing experiences of my deployment. Upon reaching the bridge, it was immediately clear we had a problem. This wasn't a sturdy bridge capable of supporting heavy vehicles. It was a rickety construction of telephone poles - not even the thicker utility poles, but the skinnier variety. The cross members were so low that our trucks would have hit their windshields trying to pass through.

I turned to the sergeants, my voice steady but firm. "You're relieved," I told them. "Return to your base. I'll take it from here and find a suitable crossing."

With our meager navigation tools, I led our convoy into the town of Ramadi, where I knew we could cross the river. But Ramadi wasn't built for oversized American military vehicles. As we lumbered through the narrow streets, we were like bulls in a china shop. Our trucks snagged electrical wires, brushed dangerously close to parked cars, and wreaked havoc on the town's infrastructure.

My heart was in my throat the entire time. I kept expecting a hail of gunfire to erupt at any moment. After all, we were hardly winning hearts and minds by tearing up their town. But somehow, miraculously, we made it through without incident.

We found the on-ramp to the main supply route, but our troubles were far from over. I noticed something terrifying as I pulled to the side to let our convoy of foreign national trucks onto the highway. With each passing truck, more and more earth was being displaced from under the on-ramp. It was like watching a disaster in slow motion.

I held my breath as truck after truck made the precarious journey up the crumbling ramp. In my mind's eye, I could see one of them tipping over, the road collapsing beneath its wheels. But against all odds, all 59 trucks made it safely onto the highway. I fell behind them, relief washing over me as we finally hit the open road back to Kuwait.

We had been driving for several hours when I noticed the trucks ahead of me pulling over. We were on a multi-lane highway cutting through a desolate stretch of desert. I had no idea what was happening - with no way to communicate with the foreign national drivers, I was flying blind.

Then I saw trucks that weren't part of our convoy speeding by, guns blazing. My instinct was to give chase, but I couldn't leave my convoy behind. I turned back, and that's when I saw them - the foreign national drivers, gesturing wildly and repeating one word repeatedly: "Shot! Shot! Shot!"

They led me to one of the trucks, its door riddled with bullet holes. I yanked the door open, and my heart sank. The driver was slumped over the wheel, blood seeping from his chest. My driver was hanging over my right shoulder, but my gunner was nowhere to be seen.

I helped the wounded man out of the cab, lowering him gently to the ground. His English was limited, but the fear in his eyes spoke volumes. "Big trouble, big trouble," he kept saying.

I tried to keep my voice calm and reassuring. "If you're going to get hurt with anybody, I'm the best person to be hurt with," I told him. "I'm going to take care of you." I pulled out my trauma dressing and applied it to one of the holes in his chest. When I asked my gunner for his combat dressing, he stared wide-eyed at me. He'd lost it.

Cursing under my breath, I cut my dressing in half with my knife to plug another hole in the center of the man's sternum. I was amazed he was still alive - the wound looked like it had gone through his lungs, possibly even his heart.

I managed to get him into the back of my Humvee and used the Mobile Tracking System (MTS) to send a message to our Tactical Operations Center. I gave them our coordinates and informed them of the critically wounded foreign national.

Their response chilled me to the bone: "Continue the mission."

I couldn't believe what I was reading. I sent another message emphasizing the critical nature of the patient's condition. The reply was even more disheartening: "Negative. Continue the mission. Enemy activity is too hot for medevac."

As I was desperately trying to figure out our next move, an Iraqi police vehicle pulled up. One of the officers spoke some English, and when I asked about the nearest hospital, he told me it was about 25 minutes away. They agreed to transport the wounded man.

I scribbled down as much information as possible about the police vehicle and the officers' uniforms, then turned to the injured driver. "You're going to be okay," I told him, hoping I wasn't lying.

With the immediate crisis handled, I turned my attention back to my team - only to find that my gunner and driver had disappeared. I finally located my driver, who claimed he'd been "pulling security." My gunner was in a state of shock, unresponsive with a thousand-yard stare. I got him into the back of the Humvee and tried to get him to drink some water, but he remained silent.

We reorganized quickly. I put my driver behind the wheel of the shot-up semi, got all the other drivers back in their trucks, and took the wheel of my Humvee myself. We drove straight through to the Iraq-Kuwait border, my nerves frayed and my senses on high alert the entire way.

At the border checkpoint, as my convoy was going through the immigration control point, I noticed the last vehicle wasn't moving. As I approached to investigate, two men with AK-47s ran off. I found the truck's owner's manual on the ground, a lone sandal that might have belonged to the driver, and discovered that the trailer's brakes had been disconnected.

After reconnecting the brakes and moving the truck through the checkpoint, I managed to locate the driver. He explained, as best he could, that the armed men had forced him out of the truck at gunpoint. He escaped and caught a ride with another truck in our convoy.

Finally, after an eternity, we returned to Camp Arifjan. At this point, I released the drivers to their supervisors, my body running on pure adrenaline and willpower.

Months later, I received news that brought a rare smile. The truck driver who had been shot not only survived but was well enough to attend a celebration. The Al-Hamadi Trucking Company was marking its first anniversary and contract renewal with the U.S. Army. At the feast thrown for Army officers, out of all the "bald white guys in civilian clothes," the driver recognized me. He came up, hugged me, and thanked me for saving his life.

He was there as one of two civilian truck drivers injured in the line of duty, and our Transportation Command (TRANSCOM) General personally awarded him a challenge coin in recognition of his service to the U.S. Army. It was a moment of light in a very dark time.

The aftermath of this mission left its mark on all of us. My gunner, still shaken by the events, was never sent out on another mission. He was assigned to a permanent night shift in our operations center, answering phones and manning the radio - we called it CQ (Charge of Quarters) duty.

I had a frank discussion with my driver about his disappearance during the crisis. While I didn't outright accuse him of cowardice, I clarified that his actions—or lack thereof—were unacceptable. He should have been by my side, assisting the injured driver, not hiding in some covered position while I was exposed on the roadside.

I was awarded the Bronze Star with Valor for my actions during this harrowing mission. The TRANSCOM General presented it to me at the end of our deployment, just before we returned to America. It was a bittersweet moment - recognition for doing my duty but a reminder of the cost and danger we faced daily in service.

The citation read:

Bronze Star with Valor Citation

The United States of America

To All Who Shall See These Presents, Greetings:

This is to certify that the President of the United States of America, authorized by Executive Order 24, August 1962, has awarded the Bronze Star Medal with Valor Device to

Second Lieutenant Roy G. Taylor, 1450th Transportation Company

For meritorious achievement and gallantry during a deliberate enemy attack while serving as convoy commander for 59 vehicles on 20 April 2004 in Southern Iraq. Second Lieutenant Taylor's tactical and technical skills and immediate assessment of the situation enabled him to render life-saving first aid to a multiple gunshot wound victim. Second Lieutenant Taylor continued to manage his convoy while responding to an enemy attack and controlled threat.

Second Lieutenant Taylor's leadership and decisiveness profoundly impacted all soldiers in his presence. His extraordinary actions reflect great credit upon himself, the 143rd Transportation Command, and the United States Army.

Awarded on 20 April 2004

Given under my hand in the city of Washington, this 26th day of October 2004

Paul E. Mock
Major General,
Commanding General
377th Transportation Corps Forward

Taylor Mess Dress

As I stood there, the weight of the medal on my chest, I couldn't help but think of all we'd been through—the close calls, the moments of terror, and the small victories. Each experience had shaped, hardened, and, in some ways, changed us forever.

The Bronze Star is among the highest awards in the United States military. Having the Valor device placed upon it elevates it even more. Valor means that I could perform my duties and keep people safe in the face of danger. This citation embodies everything I used from Boy Scouts, Civil Air Patrol, and my law enforcement experience to manage this very hectic situation.

But there was still work to complete and more missions before we could truly call this deployment over. Little did I know that our last mission would prove to be yet another test of our resolve and ingenuity. It is often believed that the final mission before rotating home was frequently the most dangerous one.

Chapter 16

Sweet America!

The journey back to America was not as straightforward as we had hoped. Around the middle of March 2005, our unit, the 1450th Transportation Company, received orders extending our deployment indefinitely. Typically, overseas rotations lasted 12 months, not including the mobilization time.

Our extension was due to the US Army's preparation for a large offensive they called "The Surge." They were keeping units like ours in theater while rotating in fresh troops. The atmosphere was tense - we'd been away from home for over 15 months, and now the finish line was moving further away.

What turned out to be my final mission happened a few weeks after this announcement. It was a large convoy to the logistics base Anaconda in Northern Iraq, one of the largest and farthest bases from our location in Kuwait. Our mission was to deliver repair parts and other essential items needed for resupply. In return, we would bring back equipment for a unit being redeployed back to the US.

As we prepared for the mission, that old childhood nightmare resurfaced in my mind again—the one of dying in a sandy environment while bleeding out. I'd long suspected it was some premonition of a wrong military mission. Now, with the stakes higher than ever, that dream felt uncomfortably close to reality.

As our convoy set out the next day, the tension among my soldiers was visible. We'd all heard the stories—the last mission before rotating home was often the most dangerous. It was as if the war saved its cruelest tricks for those on the brink of safety. This mission had me on edge, not just because of that common belief but also because of that recurring dream.

True to form, the mission was wrought with enemy attacks on our convoy. The first occurred near Taji, where two 55-gallon drums filled with explosives ignited as our lead vehicle went past. I was in the command and control vehicle, second in line. The first vehicle, a five-ton deuce and a half that had been up-armored took the blast's brunt.

Fortunately, its higher ground clearance and heavier armor plating than my vehicle likely saved lives that day. The blast wasn't as devastating as it could have been, but it left our soldiers with concussions, hearing loss, and cuts from shattered glass. Despite this attack, we pushed through the blast zone, weathering small-arms fire. We continued our mission north and rested overnight along the way.

The next day brought another close call. As we were traveling down the road, a large IED exploded beneath the pavement ahead of us, leaving a gaping hole in the road. Thankfully, we were several hundred yards away when the explosion occurred and could drive around the large hole and continue the mission.

Finally, we reached Anaconda. As we waited at the security gate to be allowed onto the post, mortar fire erupted. The crude mortars the insurgents were using didn't hit exactly where they wanted, so they would adjust their elevation and windage and fire another round.

Each shot was growing closer and closer. I found myself yelling at the MP at the gate, telling him to let us on post immediately as we were being shelled. They opened the gates, and we quickly rushed inside, grateful for another narrow escape. We completed our mission, offloading our cargo and loading up the return equipment for the port of Kuwait.

The following day, we set out early for the return trip. It wasn't until our first refueling stop that we learned of a tragedy that had occurred in the chow hall during breakfast – a suicide bomber dressed as an Iraqi police officer had detonated explosives among the service members who were eating, killing many. We had narrowly avoided this devastating attack by starting our journey earlier that morning.

Taylor in South Korea

Some months later, after my release from active duty, I learned of another attack. During that time, I was working security at a nightclub alongside an off-duty Raleigh police officer who, as it turned out, was also a military policeman with the Army Reserve. He had been stationed at Anaconda and shared a tragic story about his unit's return home.

His unit was so large that they had to split their departure over two days. The first half left without incident, but for those left behind, disaster struck. Two soldiers, mere hours away from boarding their flight home, decided to visit the Post Exchange (PX) for snacks for the journey. While waiting for their friends outside, sitting on the steps of the PX, they were killed by a mortar strike - the same kind of attack we had narrowly avoided.

The first half of the unit, still airborne, didn't learn of their comrades' deaths until they landed in the United States. This story hammered home the cruel reality of war - that danger persists until the very last moment and sometimes beyond. It reinforced the importance of constant vigilance and adherence to proper tactics and techniques throughout the deployment. Even with all precautions, making it home alive is not guaranteed.

After refueling that day, the rest of the journey back to Kuwait was mercifully uneventful. As we returned to Camp Carolina Crossroads and began the process of preparing for our return to the United States, I felt a mix of relief and lingering anxiety. The dream of my death in the sand still haunted me, and I knew I wouldn't truly feel safe until we were airborne, flying over the ocean toward home.

Weeks passed as we packed containers, loaded trucks and trailers onto rail cars, and tracked down equipment left behind due to mechanical issues or damage. My responsibility was to locate all our missing trailers – a tedious task that took several weeks but was ultimately successful.

Finally, the day came when we boarded a Jumbo 747 for the flight home. As the plane lifted off and the sandy landscape of Kuwait faded from view, I felt the tension of the past year begin to melt away. We had made it. We were going home.

As we touched down at Fort Bragg in April 2005, the air was thick with excitement, but we couldn't rush home yet. Out-processing stood between us and our families - a necessary evil of military life. We spent days reviewing personnel records, enduring medical examinations, and completing psychological questionnaires. These evaluations were crucial and designed to catch any signs of post-traumatic stress disorder or other deployment-related ailments. It reminded us that even though we'd left the war zone, the war hadn't necessarily left us.

Finally, the day arrived when my wife, Lisa, and our two kids, Vickie and Marshall, could pick me up. Seeing their faces after so long was indescribable - a flood of emotions I hadn't allowed myself to feel for months. As we drove away from Fort Bragg, I couldn't wait to get behind my special edition Ford Crown Victoria for the drive from Fayetteville back to Durham, North Carolina. That car represented normalcy, a slice of civilian life I'd been craving.

But transitioning back to "normal" life isn't as simple as changing out of fatigues and jeans. The vigilance that kept us alive in Iraq doesn't have an off switch. I found myself struggling to sleep, my body still on high alert. Crowded places made me uneasy - my mind constantly searching for threats that weren't there. My wife noticed changes, too. She told me I seemed more short-tempered and more guarded than before. It's strange how you can change without realizing it, but war has a way of rewiring you, whether you want it to or not.

Eager to return to work, I notified my employer of my return, planning to take a couple of weeks off for family time before jumping back in at the start of a pay period in May. But life had other plans. A letter informed me that my position had been eliminated during my deployment. There was no job for me to come back to.

This news hit hard. I knew from staying in touch with coworkers that they'd decided not to backfill my position while I was deployed. Instead, they'd delegated my duties to a new employee they'd hired just before I left - an ergonomist. This guy, whose job was to check workspace ergonomics to prevent issues like carpal tunnel, was now handling my emergency management responsibilities on top of his regular duties.

I wasn't about to take this lying down. I wrote back, reminding them they were obligated to return me to my position under federal law. I pointed out that I had a master's degree in occupational safety, while the ergonomist only had a bachelor's. My five-year tenure with the company and superior qualifications should have meant that he would be laid off and I'd take over the combined role.

But the company dug in its heels. They insisted my position no longer existed and I had no role to return to. Left with no choice, I filed a complaint with the Federal Department of Labor under the USERRA law—the Uniformed Services Employment and Reemployment Rights Act. This law mandates that companies reinstate soldiers returning from deployment at the same pay rate and in the same or a similar position to the one they left.

The process was frustrating. I spoke with a caseworker who believed I had a strong claim. They contacted my employer to inform them that their actions violated federal law. But the company remained obstinate, seemingly willing to fight the law. What I discovered later added to my frustration—the law lacked teeth. There were only real consequences for businesses that relied on federal government grants or contracts for violating USERRA. It was a federal law without punishment, leaving returning service members like myself in a precarious position.

The negotiations dragged on for six months, after which the company finally offered one year's salary to end the complaint. After discussions with my caseworker, who was becoming overwhelmed with similar cases as more soldiers returned from Iraq and Afghanistan, I reluctantly accepted the offer.

This experience left a bitter taste in my mouth. I stayed in touch with my co-workers during my deployment, even submitting photographs and stories for the company newsletter. Upon returning, being treated this way felt like a betrayal not just of me but of every service member who puts their life on hold to serve their country.

Reflecting on these experiences, I'm reminded daily of the sacrifices made by the men and women who serve in our active military, Reserve, and Guard. They protect our country and ensure the freedoms we hold so dear. With recruiting numbers down, finding people willing to step up and be part of the 1% in military service is becoming increasingly difficult.

The need for dedicated public servants has never been greater. If you're feeling a call to serve, I urge you to consider answering that call. The world needs more individuals willing to put on the uniform, pursue knowledge relentlessly, and dedicate themselves to making their communities safer and stronger.

Although the challenges are real, so also are the rewards. You'll be pushed to your limits but discover strengths you never knew you had. You'll face dangers but also experience a sense of purpose and camaraderie that's hard to find anywhere else.

Choosing this path will set you apart. But as I'd say, it's not about being better than others – it's about pushing yourself to be the best version of yourself and using that growth to serve your community and your country.

Remember, the American Dream isn't handed to you – it's earned through hard work, perseverance, and a willingness to serve something greater than yourself. Whether you choose to serve in the military, pursue a career in law enforcement, or dedicate yourself to academic excellence, know that you're embarking on a path that will challenge, change, and, ultimately, allow you to make a real difference in the world.

During that period, I continued pushing on my job search, applying for police chief positions in the surrounding area and other exciting roles in the civilian market. Eventually, an opportunity arose that would shape the next phase of my career - an interview for the position of Chief of Police for the National Geospatial-Intelligence Agency in Washington, DC. I accepted the offer, marking my path back to law enforcement.

This return to civilian life was a gradual process. It took about six months after returning from Iraq before I felt truly settled. My nervousness in public spaces began to subside, and I slept better through the night. My family relationships, strained by the changes in my demeanor, also began to improve.

During those early days, most of my energy was focused on building my private police business, Capitol Special Police. The business was growing rapidly, and I found myself hiring new officers, expanding our coverage area, and working many shifts myself to sustain our growth. It was a period of significant investment which involved purchasing more police vehicles, uniforms, and firearms to keep up with our increasing number of contracts.

Before all this, one surprise awaited me when I first returned home to Durham. My wife had stored all the things I'd bought over the internet while deployed in a spare room of our house. She'd never opened them, leaving me to return to what felt like a mega Christmas. At least a quarter of the room was filled with boxes of things I'd purchased - some of which I didn't even remember ordering.

This abundance of packages was a strange side effect of deployment shopping. Everything went to our home address since I couldn't have items shipped to me in Iraq. Without the immediate gratification of receiving and using these purchases, the desire to buy more never disappeared, so I'd order more than I typically would under normal circumstances.

Opening those boxes was an adventure in itself. There were radio-controlled helicopters, model cars, antique police paraphernalia—a wide array of items that had caught my interest during late-night internet browsing sessions in Iraq. She had never even mentioned when these packages arrived, making the whole experience even more surreal.

But amidst the challenges of reintegration and job uncertainty, there was also time for much-needed relaxation and family bonding. We had planned a seven-day Caribbean cruise to spend quality time together as a family. It was a relief to be out of my military uniform and in average temperatures, a far cry from the 150-degree heat we endured in Iraq.

We loaded the car and drove to Cape Canaveral, Florida, to board a Carnival ship bound for the Bahamas. We had also planned several excursions that the kids were excited about. One of these excursions was particularly special for my daughter, Vicki. She had always dreamed of becoming a veterinarian to save animals, and we'd arranged for her to swim with dolphins. Seeing her face light up as she interacted with these intelligent creatures was a moment I'll never forget - a reminder of the simple joys of family life that we often take for granted.

We had other adventures, too - lunch on a private island, shopping in local markets, and exploring the vibrant culture of the Bahamas. It was a much-needed opportunity to relax without worrying about the daily grind or the constant vigilance required in a hostile environment.

Our family time wasn't limited to the cruise. We also visited other attractions along the way. I fondly remember us eating at Bubba Gump's restaurant, laughing over shrimp and movie quotes. We also toured a US aircraft carrier on display in Charleston, South Carolina - a different perspective on military life for the kids, seeing the massive scale of naval operations compared to my Army experience.

Grandpa Taylor and grandchildren

This vacation was more than just a trip—it was a chance to reconnect, talk to the kids about my deployment, and catch up on their lives. We shared stories, made new memories, and began knitting our family back together after so much time apart.

I felt a renewed sense of purpose and connection as we returned home from our cruise. The challenges of reintegration were still there—the job search, the lingering effects of deployment, the adjustment to civilian life—but this time with my family reminded me of what I'd been fighting for. It wasn't just about serving my country; it was about protecting the freedom and safety of families like mine, ensuring they could enjoy simple pleasures like a family vacation without fear.

Chapter 17

Intelligence Agency

The call came on a crisp autumn morning. I had been selected as the first police chief of the National Geospatial-Intelligence Agency (NGA). This wasn't just any law enforcement position – it was a role at the forefront of national security, tasked with protecting some of the most sensitive information and facilities in the United States.

As I drove through the gates of our headquarters in Bethesda, Maryland, I couldn't help but feel a sense of awe. The NGA, formerly the Army Mapping Agency, had evolved far beyond its World War II origins. While its primary mission remained to produce maps for all U.S. military services, the agency had taken on responsibilities crucial to national security in the post-9/11 world.

"Everything about the agency, other than its location and name, is pretty much classified," I'd tell friends and family when they asked about my new job. It was an understatement; the NGA had become a linchpin in the nation's defense infrastructure, monitoring vessels worldwide for distress signals and coordinating emergency responses. But our role after September 11th, 2001, underscored the gravity of our mission.

Our main facility in Bethesda was the highest-altitude building in the D.C. area. In the wake of the 9/11 attacks, we became the fire control center for all surface-to-air missile batteries stationed around the National Capital Region. The roof of our building housed a radar monitoring station with firing controls for these missiles, providing visual and instrumented radar information to protect our capital from another airborne attack.

But perhaps our work's most critical—and classified—aspect involved satellite feeds. We took raw data from the National Reconnaissance Organization (NRO) and transformed it into actionable intelligence for other three-letter agencies across the U.S. government. This responsibility weighed heavily on all of us.

You might wonder why they needed a police chief for such a specialized agency. Well, the threat landscape had changed dramatically post-9/11. The previous Department of Defense security force was no longer sufficient. They needed a police force with expanded authorities, and that's where I came in.

During my hiring interviews, I was told something exciting and challenging. The agency representative leaned forward, his eyes intense, and said, "We want to have the most prestigious law enforcement agency in the U.S. government. We're aiming for national accreditation standards and to excel in every aspect of our operations. Can you make that happen?"

I remember sitting up straighter in my chair, meeting his gaze. "Absolutely," I replied. "I'm the right man for the job." And I meant it.

After I was appointed chief, I learned another interesting tidbit about why I was selected. Besides my law enforcement leadership background, my master's degree in occupational safety played a role. The agency believed this knowledge would be crucial in developing safety protocols in case of another attack like the one on the Pentagon. It was a sobering reminder of the world we now live in.

After getting oriented, my first order of business was to start working on the policy manual. I focus on creating policies for high-liability areas: use of force, vehicle pursuits, firearms qualification—you name it. If not handled properly, these areas can be real thorns in law enforcement agencies.

I hit an unexpected snag when I presented these policies to my direct supervisors and all career CIA employees in their security division. To them, "security" meant information security, not physical security. They didn't have a solid grasp of law enforcement or physical security protocols for keeping everyone safe.

I remember being asked, "Why do we need a pursuit policy?" The puzzled look on their faces was genuine. From their perspective, it was a fair question, so I laid out some scenarios for them.

"Let me give you an example," I said. "Just recently, we had a domestic dispute that resulted in a stabbing at work. A boyfriend and girlfriend who worked here got into an argument, which escalated quickly. The man stabbed the woman right here on our premises."

I could see their eyes widening as I continued, "Now, what if that man had run, gotten to his car, and made it out the gate before we could catch him? Would we be allowed to pursue him?"

They looked at each other, then back at me. "I... don't know," one of them admitted.

I nodded. "Exactly. That's why we need these policies. We have over one hundred police officers; they need to know what they can and can't do in these situations."

I gave them another scenario: "Imagine an employee has some mental health crisis. They set the ladies' locker room on fire, then go out to the parking lot and start doing donuts in their car right in front of our main building."

I paused for effect, then asked, "What if they tried to flee through the gate? Could we chase them?"

Again, they were stumped. I could see the gears turning in their heads as they considered the implications.

My final scenario hit closer to home for them. "Imagine someone took classified information and was leaving the facility," I said. "We get word from their supervisor just as they pass through the gate. Can we chase them to retrieve the classified documents?"

Once more, they didn't know. Momentarily, the room was silent as the gravity of these potential situations sank in.

The need for a comprehensive use of force policy became evident as I explained our officers' arsenal: Beretta 92F nine-millimeter semi-automatic pistols, MP5 machine guns, and Remington 870 12-gauge shotguns. This was serious hardware for a serious job. I was shocked to learn that in the past, the agency had never qualified with all its weapons at nighttime; it was only qualified during daylight hours.

I explained that we worked 24 hours daily, and all our facilities were in densely populated urban areas. I emphasized that we couldn't risk a stray bullet striking someone or causing property damage, hence the need for nighttime qualifications.

While my supervisors understood these reasons and appreciated my work, most policies were never officially approved during my tenure. The only policy that received the stamp of approval was one I didn't even write - the uniform loan policy. It stated that if an officer soiled or damaged their uniform while working, our quartermaster could issue them a temporary replacement until theirs could be repaired or cleaned.

The disconnect between the agency's initial ambitions and what they eventually approved was striking. But despite this, the NGA remained an incredible place to work.

Working there was unlike any other law enforcement job I'd experienced. The facility itself was a marvel of security measures. Every room in the building, save for the first-floor bathrooms and cafeteria, was a Sensitive Compartmentalized Information Facility (SCIF). Access required a proximity badge with the proper and physical code similar to a safe's door.

While my officers had access to every room for emergencies and patrols, we respected the sensitive nature of the work and only entered when necessary. The police officer's primary duties were to act as a visible deterrent against criminal activity and to guarantee that only authorized personnel had access to the base and its facilities.

Air Force Lieutenant General James Clapper, a well-respected figure in the intelligence community, commanded our unit. The hallways often buzzed with activity as high-ranking officials like Secretary of Defense Donald Rumsfeld and Secretary of State Colin Powell visited for briefings. Seeing figures I'd only ever watched walking through our facility on television news was surreal.

When General Clapper announced his retirement, we began planning a farewell gala to include President George W. Bush and most of his cabinet members. Coordinating security for an event involving the sitting president and top government officials was one of the most intense tasks I have ever carried out. We also planned to erect three circus tents in front of our building on the ellipse, complete with bands and catering.

As the date approached, we learned that the president had taken most of his cabinet on an unannounced trip to Iraq to serve Thanksgiving meals to the troops. While disappointing, the show continued, and we ensured the event was a fitting tribute to General Clapper's service.

General Clapper's successor was Vice Admiral Moret, a career Navy officer. I got to know him better than I had General Clapper, primarily due to Admiral Moret's love for physical fitness. He often led weekly runs with all the military personnel in the facility, and I would run alongside him.

"Chief," he'd call out to me as we started our run, "you're my personal bodyguard today!" It became a running joke between us. As we jogged along the trails around Bethesda, we'd have great conversations about everything from national security to staying fit.

Those runs gave me a profound sense of national pride. I'd sometimes extend my route, running from our facility down to the heart of D.C., circling the monuments, and taking in the sights and sounds of our nation's capital. The sound of military bands practicing, the sight of the gleaming white monuments against the blue sky – it was an excellent time to be an American, working in the heart of our government.

During this period, I also transferred from the North Carolina National Guard to the Virginia National Guard, serving as the executive officer for the Joint Force Headquarters. During a weekend drill, the J3 (Joint Operations Officer) approached me about a full-time provost marshal position.

"Roy," she said, pulling me aside after a briefing, "there's a full-time provost marshal position open. I think you'd be perfect for it." She explained that the role involved overseeing all law enforcement, security, anti-terrorism, and operational security programs for the military.

The position was primarily based at Fort Pickett in Blackstone, Virginia, about an hour north of the North Carolina border. When I looked into it, I found that the salary matched what I earned at NGA. I could make a lateral transfer since both agencies fell under the Department of Defense umbrella. The clincher was the much lower cost of living in southern Virginia compared to the D.C. area.

After careful consideration, I decided to accept the position. When I gave my two weeks' notice to NGA, it was with mixed emotions. Looking back on my year there, I felt proud of our accomplishments. Even though not all the policies were fully implemented as I'd hoped, we'd made significant strides in training our officers, ensuring proper weapons qualifications, and elevating the overall professionalism of the agency.

Chapter 18

From Guard To Reserve

When I joined the National Guard in 2000, I had no idea what was in store. I just knew I wanted to serve my community. Looking back, that's where it all begins—starting right where you are and giving your best, even when you don't fully know where the path will lead. It's not always about having a grand plan; it's about showing up and doing what's needed, whether responding to a local emergency or stepping into something entirely new.

With a desire for service and availability to serve, my journey led me from the young recruiter's office to the Iraqi desert. It also exposed me to some of our time's most critical national security challenges.

But here's the thing: you can't stay in one place forever. At some point, you need to grow, learn, and push yourself to expand. Service is essential, as is stretching and exercising to make an even more significant impact. Each one prepared me for something bigger and more meaningful as I moved through different roles and challenges. It's a process of constantly evolving, adding new skills, and facing bigger challenges so you can serve in new and greater ways.

My full-time career with the Virginia National Guard began when I was appointed their new Provost Marshal, or as we called it in the joint world, the J34. In the military, "joint" refers to operations involving multiple service branches working together. In my new role, I collaborated with members of the Army, Navy, Air Force, Coast Guard—and occasionally Marines—in our joint units.

The primary focus of my position was anti-terrorism, which meant taking proactive steps to prevent possible terrorist attacks. This role placed me on the Adjutant General's staff, directing me to the top. At the time, our Adjutant General was Major General Robert Newman, an Air Force general known for his personable nature and a serious commitment to the safety of his troops.

I'll never forget what General Newman said when I first assumed the role. "Roy," he began, his voice carrying the weight of command, "the thing that keeps me up at night is somebody dying on my watch." He looked me straight in the eye and continued, "I want you to make sure that a terrorist attack kills none of our soldiers or airmen."

Those words became my mission statement. This was in 2005, several years after 9/11, but with the wars in Iraq and Afghanistan still raging, the threat to the homeland remained very real. The gravity of my responsibility hit home at that moment.

Taylor and Inspection Team

One of my primary responsibilities was traveling around the Commonwealth of Virginia, inspecting our readiness centers (what people would call armories), maintenance facilities, aircraft hangars, and two major installations: Fort Pickett in Blackstone and the State Military Reservation in Virginia Beach.

I approached this task systematically, starting at one end of the state and working my way to the opposite end. It was time-consuming, but the safety of our service members was worth every minute. I inspected 54 readiness centers, scrutinizing their anti-terrorism programs, physical security measures, operational security protocols, and personnel security practices.

To my dismay, most units failed their initial inspections. I found numerous deficiencies across all areas of security. Each week, I reported these failures to the general. One day, curiosity got the better of him.

"Roy," he asked, a hint of disbelief in his voice, "are you breaking into these facilities? How do you get in?"

I couldn't help but smile as I explained my method. "No, sir. I conduct surveillance from across the street and watch where the smokers go. Typically, they use one door and block it open with a small rock or other object to keep it from locking behind them. Most of the time, they forget to take the rock with them when they're finished. That's my primary way to get into the facilities."

The General's eyebrows shot up. "I don't believe this," he said, shaking his head. "I want to accompany you on one of your inspection tours."

We set a date, and when the day came, the plan was changed. Instead of flying out, we decided to check the armory in Richmond. He insisted we meet by the motor pool so no one would recognize his vehicle.

As we approached the motor pool, the General pointed to the chain-link gate. "Well, the motor pool is secure," he said confidently.

I couldn't help but chuckle. "No, sir, it is not."

He looked at me, puzzled. "What do you mean? It's got a chain and a lock."

I explained, "The U.S. Army does not use Master locks for securing anything." To demonstrate, I grabbed the padlock with both hands and violently snapped it downward, causing it to break open. "That's how easy it is. We're inside the motor pool now."

The General's eyes widened. I had brought two of my subordinate NCOs to assist. I instructed one to check all the vehicles in the motor pool for proper security, and the other was tasked with checking exterior doors leading into the armory from the motor pool.

We were inside the armory within minutes through an unlocked supply room door. We wandered the halls, and to the General's growing frustration, no soldier acknowledged the General's presence or addressed him adequately.

Finally, frustrated, the General said, "Let's go see the commanding officer."

When we burst into the commanding officer's office, the poor colonel stood up so fast his chair flipped backward. Suddenly, it was, "Good morning, General! How may I help you, sir?"

What followed was a dressing-down for the ages. The General berated the Colonel for the security lapses and his troops' failure to render proper courtesies. He ordered the entire brigade staff to report to his office the following Saturday with a comprehensive plan to address their security concerns.

That Saturday, the briefing room was packed. The unit had prepared an elaborate PowerPoint presentation outlining their failures and proposed solutions. The General let them speak for about two minutes before cutting them off.

"Gentlemen, that's sufficient," he said, his voice steely. "My message to you is this: I don't want anyone to die on my watch. Follow the regulations. Make sure your readiness centers are properly secured. Follow all protocols so that everybody remains safe." He paused, letting his words sink in. "You go back and tell them that General Newman chewed your ass and said it better not ever happen again."

That's the commitment and insight necessary to be a good leader. He didn't berate them harshly but clarified that future failures would not be tolerated. That incident sent shockwaves through the Virginia National Guard. Word spread quickly that the General was serious about physical security and anti-terrorism programs.

My ability to access armories undetected dwindled drastically after that. It became rare to find the same level of failures we had encountered before the General's intervention.

As time passed, my responsibility expanded even further with the addition of emergency management and consequence management programs. This required me to deal with police, fire, and EMS response on our bases. State employees staffed Fort Pickett's police department. It was technically a state agency because the National Guard, when not federalized, is an asset of the state's governor.

We had our own police department, dispatchers, and MP units that provided security and law enforcement services on their drill weekends. This additional requirement allowed me to add one more person to my staff to assist with the program.

It was a large undertaking because national requirements mandated a 911 and other systems typically found on larger installations. Implementing these at readiness centers and smaller operations proved challenging. We had to coordinate closely with local public service agencies to ensure they could respond to our armories, maintenance facilities, aircraft facilities, and installations as necessary.

One of the most interesting assignments I received during this time was leading an investigation into the misuse of a recruiting program called G-RAP (Guard Recruiting Assistance Program). This initiative allowed National Guard members to bring friends or family to join the Guard. The recruiter would receive $1,000 when the new member signed their contract and completed testing and physical requirements, then another $1,000 when they are shipped out for basic and advanced training.

It sounded great in theory, but the program lacked adequate safeguards. People started taking advantage, falsifying information, and disregarding rules. The government's losses cost hundreds of thousands of dollars across the United States.

The U.S. Army's Criminal Investigative Division (CID) then launched what they called a "RAPTOR" investigation. As part of the National Guard's contribution, we were assigned several cases to investigate with the understanding that if criminal suspects were identified, CID would adopt those cases.

Brigadier General Janice Igou selected me to lead the investigation for the Virginia National Guard. When she laid out her initial idea of how the investigation should be conducted, I respectfully disagreed and presented my investigative plan. She was impressed with my quick thinking and gave me the green light.

I started by interviewing soldiers who had joined through the G-RAP program, identifying the "recruiters" who had recommended them and received the $2,000 bonuses. One sergeant's name kept coming up, but the new soldiers didn't know him personally. They only knew his older brother, a senior recruiter for the Virginia National Guard.

I brought the sergeant in for questioning and read him his military version of Miranda rights. He quickly ended the interview and left my office. I knew exactly where he was headed—straight to his older brother, which became my next stop.

I introduced myself to the recruiter, explained the purpose of my visit, and read him his Miranda warning. To my surprise, he was very open and honest, making a full confession about his and his brother's involvement in the scam.

The scheme worked like this: When recruits came to the recruiter's office (the older brother's full-time job), he would give his younger brother all their personal information before submitting the paperwork. The younger brother would then enter this information into the G-RAP system, making it appear that he had recruited these individuals. Once the information was uploaded, the older brother - the recruiter - would enlist the recruit, and his younger brother would receive the initial entry bonus and the follow-up payment after the recruit completed their training.

The recruiting sergeant claimed he never took any money, insisting he only tried to help his brother, who was facing financial difficulties. He also provided names of other recruiters engaged in similar practices with the G-RAP program. This information quickly escalated the case, and the Army CID adopted it. But I wasn't done yet.

My next target was a Lieutenant Colonel who worked full-time for the National Ground Intelligence Center (NGIC) as part of his National Guard duties. He was also an ROTC (Reserve Officers' Training Corps) instructor at a college in Richmond. According to reports, he had told his class that anyone interested in joining the National Guard should go through him so he could get money through the G-RAP program - money he claimed to use for family vacations.

I began by investigating all the new soldiers who had come through the program, listing him as their initial recruiter. When interviewed, they all said the same thing: the Colonel never monitored or mentored them through the process, a requirement of the Wrap Back program. Many confirmed that he had mentioned using the money for family vacations.

I scheduled an interview with the Colonel. At the time, I was still a captain. I explained the circumstances of my investigation and read him his Miranda rights. He claimed ignorance when I asked about the requirement to mentor the soldiers he had entered into the program.

I produced a certificate showing he had completed an online training course explaining the program's rules. He had signed it, acknowledging he would abide by all regulations.

"Do you remember that?" I asked.

"No," he replied.

"I understand that it took you one minute and 38 seconds to complete this PowerPoint program and sign off, acknowledging that you would abide by all the rules of this program," I pressed.

Faced with this evidence, he admitted that he had clicked through the program without reading the instructions and rules.

I explained that he was now the subject of a criminal investigation and that the Pentagon would likely suspend or revoke his top-secret security clearance.

Given his rank and position at NGIC, I contacted the general and explained that I was obligated to report this criminal investigation to the Pentagon's security office. Failure to do so would result in the immediate revocation of his top-secret security clearance. The general gave me the go-ahead.

She then contacted our JAG officer (Judge Advocate General - the military's attorneys). The JAG officer reviewed my report and the relevant regulations, agreeing with my assessment and confirming my responsibility to report.

I took no pleasure in exposing a well-meaning officer for this criminal activity, but it was a clear violation, and everyone had to be treated equally. The next day, all of the Colonel's accesses were revoked, and his personal belongings were gathered and placed in a box at the NGIC's office in Charlottesville, Virginia. It marked the end of his full-time employment with that agency.

These investigations were challenging and often disheartening. It's never easy to uncover wrongdoing among your fellow service members. However, it highlights how greed and lax oversight can lead even well-meaning individuals astray. One of our Army values is integrity, and this case demonstrated the consequences of failing to uphold that standard.

Another difficult assignment I was given, based on my experience in law enforcement and as the Provost Marshal, was investigating the suicides of service members. Typically, these investigations would occur several days after the incident, after local law enforcement had handled the initial response. My role was to review the files, manner of death, and other relevant circumstances.

Contrary to what many might assume, post-traumatic stress disorder (PTSD) from military deployments wasn't the primary cause in the cases I investigated. The last six investigations I conducted before leaving the National Guard for the Army Reserve involved young males experiencing domestic difficulties with their wives or girlfriends. These individuals had taken their own lives while off duty from the National Guard.

None of the six soldiers had been deployed, and none were active duty or full-time National Guard soldiers. They were all called "M-day," meaning they were traditional guardsmen who drilled one weekend a month and two weeks a year for annual training.

Tragically, no warning signs were exhibited, according to their first-line supervisors or any of the command staff in their assigned units. It's heartbreaking to see people driven to suicide because of personal relationship issues.

I recall another case of a soldier who kept wrecking his car regularly. I finally determined he was trying to kill himself in a car accident. We were able to get him committed to treatment, and as far as I know, he's still alive today. Unfortunately, in most of my cases, the individuals were successful in their suicide attempts, leaving behind grieving loved ones and unanswered questions.

These investigations occurred at a time when the military was under intense public scrutiny regarding suicides. The National Guard took the issue very seriously, but in these cases, there were no signs of mental distress during the weekends when their sergeants had interacted with them. It was a reminder of the complex nature of mental health issues and the importance of support systems both within and outside the military structure.

As my tenure with the Virginia National Guard drew close in May 2017, I faced a crossroads. My wife and I had decided to move to Newberry, South Carolina, which would put me about five and a half hours away from my post in Blackstone, Virginia. The monthly commute seemed daunting, so I began exploring options closer to my new home.

Initially, I contacted the South Carolina National Guard, hoping to find a position there. When that didn't pan out, I cast a wider net, applying to the Army Reserve. Within hours, I had multiple offers on the table. My heart was set on Fort Bragg, two and a half hours from Newberry. But as luck would have it, that position had just been filled.

Instead, I was offered a role with the 346th MP Company in Nashville, Tennessee. The irony wasn't lost on me—I'd traded one five-and-a-half-hour commute for another and was entering an unfamiliar organization with no existing relationships. But there was a silver lining. The drive to Nashville took me through Knoxville, where my mother was in a rest home. Visiting her regularly on my way to and from drill seemed like an opportunity.

Sadly, I only had two good visits with her during my first month in the new assignment. The following month, her condition had deteriorated significantly, and she was asleep when I stopped by. The month after that, as I drove to drill, I received the call every child dreads. My brother informed me that our mother had passed away just eleven months after our father's death.

Fay and Marie Taylor were not just my adoptive parents; they were the bedrock upon which I built my life. Their unwavering commitment to service, deep-rooted faith, and quiet strength shaped the man I became. It wasn't just the love they gave me that left a lasting mark, but the way they lived their lives—always thinking of others and ready to help no matter the cost.

Their passing, within a year of each other, felt like the closing of a chapter. Yet, their influence continues to echo in every decision I make. With her gentle spirit, my mother taught me the power of compassion, while my father, through his discipline and resolve, instilled a relentless work ethic in me. Together, they showed me what it means to live a life of purpose.

It's hard to sum up the impact of two people who gave so selflessly, but perhaps it is enough to say that they left the world better than they found it. From the children they adopted and nurtured to the countless lives they touched in their community, Fay and Marie embodied what it means to serve others. Their legacy isn't one of material wealth or grand achievements but one of kindness, integrity, and love.

Their passing reminded me that life, like military service, is full of unexpected turns. But each experience builds and prepares us for challenges we never saw coming. My journey through law enforcement, emergency management, and military service had come full circle, equipping me for roles I could never have imagined as a young recruit. As I reflect on my career, I'm reminded that the path of service is rarely straight, but it's always worthwhile.

Taylor off-duty in Yongsan Korea

Chapter 19

Anti-Terrorism Unit

When I was a Provost Marshal, my experiences in law enforcement and the military converged unexpectedly. I found myself on the front lines of America's fight against terrorism, and the skills I had honed over years of service in various capacities were tested in ways I could never have imagined when I first put on a uniform.

One of the most significant roles I took in that chapter of my career was as a member of the FBI's Joint Terrorism Task Force (JTTF). In many ways, it was a position that fulfilled a lifelong dream. I had always been intrigued by becoming an FBI agent, but circumstances and age restrictions had previously thwarted those ambitions.

I vividly remember my first attempt to join the Bureau. I had just finished my bachelor's degree and was 28 years old. Full of hope and ambition, I called to inquire about the application process. The response was like a bucket of cold water: "I'm sorry, but you're too old. You cannot have reached your 28th birthday."

Several years later, when the age limit was raised to 30, I tried again. Once more, I was met with disappointment. "You cannot have reached your 30th birthday," they told me. My dreams of becoming a Federal Bureau of Investigation Special Agent seemed permanently out of reach.

But fate has a funny way of opening doors when you least expect it. My role in the National Guard and the emerging issues involving Guard members in potential terrorist activities made me a valuable asset to the

Joint Terrorism Task Force. Suddenly, I worked alongside FBI agents and other task force officers from various local and state agencies. It wasn't exactly how I had envisioned joining the FBI, but it was an opportunity to serve that I embraced wholeheartedly.

The work was intense and often sobering. We were dealing with real, credible threats to American lives on American soil. I remember one particularly chilling period when we received intelligence that Al-Qaeda was planning to kidnap a soldier and their family to behead them on television. The threat was specific to the Virginia area, and my job was to help mitigate the risk to our service members.

We immediately ordered soldiers not to wear their uniforms off base. It was a drastic step, but one we felt was necessary given the gravity of the threat. I added my caveats to the order, cautioning against displaying military occupational specialty logos or awards like paratrooper jump wings on personal vehicles. These seemingly innocent displays of pride could make our personnel and their families targets.

I tried to drive the point home to our troops: "You're out shopping at the grocery store with your wife and kids, pushing your grocery cart out in the parking lot, and suddenly you're kidnapped by terrorists. This is the reality we're facing. We need to be vigilant at all times."

The threats we dealt with were varied and often horrifyingly creative. Another piece of credible intelligence suggested that terrorists in the area were planning to fire a shoulder-launched rocket at an Amtrak train during rush hour to maximize civilian casualties. This led to many nights spent monitoring railways during peak commute times, our eyes peeled for suspicious activity.

It was a scary time in America, and the scope of the FBI's intelligence operations was staggering. The number of people under surveillance was far greater than I had ever imagined. But given our threat level, it was a necessary evil. I felt immense pride in serving alongside the dedicated men and women who were working tirelessly to keep our country safe from these terrorist attacks.

My role extended beyond the JTTF. I was also integral to the Virginia State Police Fusion Center, an intelligence gathering and dissemination hub. My position as a full-time provost marshal gave me access to military resources that proved invaluable in these investigations. Anytime a military member was a witness, victim, or suspect in a case, I could quickly gather information and data about that soldier, helping to facilitate investigations and provide crucial context.

One of the most challenging aspects of this work was the constant state of alertness it required. The threats were not abstract concepts or distant possibilities – they were real, immediate, and potentially catastrophic. Every day, they brought new intelligence, threats to assess, and risks to mitigate. It was a relentless cycle that demanded unwavering focus and dedication.

But it wasn't just about reacting to threats. Much of our work involved proactive measures to make our military installations and personnel less vulnerable to attack. This is where my experience as provost marshal came into play. I worked tirelessly to improve security protocols, tighten access controls, and educate our personnel about the importance of operational security.

As the threat landscape evolved, so did our approach to countering it. My final assignment in the National Guard, before reverting to an M-Day position, represented a shift in focus from conventional terrorist threats to the specter of nuclear, biological, and chemical attacks on U.S. soil. It was a role that required a unique blend of skills – part law enforcement, part military strategy, and part disaster response.

Taylor MG Ernst Disaster Response

After my mother's passing, I renewed my commitment to my new role with the 346th MP Company in Nashville, Tennessee. This role presented a unique challenge that tapped into my law enforcement background. Our mission was to inspect prisoner-of-war camps and military disciplinary barracks like Fort Leavenworth, as well as correctional facilities such as the one the Navy has in Charleston, South Carolina.

This assignment required a deep dive into the intricacies of correctional facility operations and the nuances of international law. I had to become very familiar with the rules and regulations governing the operation of these correctional facilities and prisoner-of-war camps, including the Geneva Convention.

The Geneva Convention, created in 1864 but revised and adopted in 1949 after World War II, outlines how prisoners of war should be treated and the obligations of each signatory nation. Our job was to ensure these standards were met, not just in theory but in practice.

Our inspection teams left no stone unturned. We scrutinized food preparation areas, ensuring meals met nutritional standards and were prepared in sanitary conditions. We reviewed medical records and facilities, verifying that prisoners received adequate care and treatment for injuries or illnesses. We even examined each facility's physical infrastructure, checking everything from electrical systems to plumbing to ensure they were in good working order and didn't pose security risks.

It was meticulous work but vital to upholding our nation's commitments and maintaining our integrity on the world stage. My performance caught the eye of the brigade commander. He was impressed with my improvements and the training opportunities I arranged for our soldiers.

For our first year's annual training, I organized an inspection of the brig in Charleston, South Carolina. This hands-on experience with a working correctional facility was invaluable for my team. We conducted a thorough inspection and presented our findings and recommendations to the facility commander.

My success in this role led to an unexpected offer. The brigade commander called me one day with a proposition. "Roy," he said, "you wanted to go to Fort Bragg. Here's your opportunity. Do you want to take over the 138th military police detachment there?"

I surprised him with my response. "No, sir," I said firmly.

"What?" he exclaimed, clearly taken aback.

I explained my reasoning. "You asked if I wanted to take it, and the answer is no. I've heard the unit has a lot of problems. The last several commanders have been relieved from their positions due to failures in the unit. I don't want to risk ruining my military record."

But I also understood the implications of refusing an assignment. "However," I continued, "I know it's not good for your military career to refuse an assignment. So no, I don't want to take it, but if that's where you believe I should serve, then yes, sir, I will accept it."

The commander seemed relieved that I wasn't outright refusing the position. He assured me, "Roy, as long as you can move the football a few hundred feet toward the goalpost, I'll be happy."

"Yes sir," I replied, "I will do the best I can."

And so began my tenure with the 138th Theater Detention Reporting Center (TDRC), the only one of its kind in the United States Army. Our mission was critical: when conflict broke out, we were responsible for correctly processing prisoners of war and civilian detainees. This meant recording all their biographical data, facilitating communication with their families, assigning them official numbers for the national archives, and ensuring any intelligence gained was channeled correctly.

When I arrived, however, the unit was in disarray. Weapons qualifications were outdated, physical fitness tests were overdue, and administrative failures abounded. As I dug deeper, I uncovered a systemic issue: recruiters funnel soldiers from active duty at Fort Bragg into our unit without even notifying me. I'd receive monthly reports showing 80 new names I'd never heard of, much less met or trained. It was an administrative nightmare.

I knew I had to act fast. My first step was to confront the recruiters. I clarified that our vacancies were for military police only, not a catch-all for every MOS (Military Occupational Specialty) under the sun. I instituted a new policy: no soldier would be accepted into the unit without my direct approval and a personal conversation.

This drastically reduced the influx of mismatched personnel. For those few who still slipped through, I tried to contact them, understand their interests, and either accommodate them or refer them to more suitable units. I noticed that many of these soldiers were returning to their hometowns and didn't fully understand the commitment of joining the Army Reserve.

Apart from the influx of mismatched personnel, equipment was another major concern. We lacked sufficient vehicles to transport our personnel, and our technology was woefully inadequate for our mission. Our job centered around entering data into specific computer programs as we processed detainees or prisoners of war. Still, we were short on the computer terminals necessary for our data entry tasks. Given the nature of our mission - requiring deployment on short notice - these deficiencies were unacceptable.

I secured an additional room in our building and worked with the base IT group to set up a computer lab. This allowed our soldiers to practice using the systems they'd need in wartime during our drill weekends. I also focused on getting our soldiers range-qualified on their weapons and coordinated with medical units to schedule physical exams during drill weekends, ensuring compliance with these critical requirements. Slowly, we began to address the unit's deficiencies.

During that time, our progress was closely monitored. We had monthly meetings with our commanding general - a two-star general who insisted on reviewing the deficiencies of each soldier. She would name by name through each unit's roster, calling out individual soldier deficiencies. It was an unusual level of micromanagement for someone at her level, but it underscored the importance of our mission and the need to maintain readiness.

Typically, this wasn't a general's responsibility—it wasn't even the commander's. Such details would be handled by first sergeants and NCOs, with commanding officers ensuring they were completed. But here we had her going down the line for every unit under her command.

These meetings were exhausting, often lasting into the early morning hours as each unit waited its turn to be scrutinized. But we persevered and made significant strides in addressing these issues. By the time I left for my next assignment, which was my last assignment with the military, the unit was in good standing.

My final role in the military brought me back to my roots in emergency management. I saw an advertisement for the state emergency preparedness liaison officer position in North Carolina. Having moved back to the Raleigh area, this opportunity aligned perfectly with my experience and location.

I approached my brigade commander about my interest in this unit, explaining my experience with emergency management, particularly my role as provost marshal in the National Guard and the numerous state emergency management missions I had completed. He encouraged me to apply and promised to release me if I got selected. I applied, and with my emergency management credentials and master's degree in occupational safety, I was the standout candidate for the job.

This new role placed me at the heart of disaster response in North Carolina. Based at the North Carolina National Guard headquarters, which houses the state's emergency operations center, I coordinated military resources during man-made or natural disasters. It was a joint position, requiring me to work across all military branches.

My duties involved attending all the governor's briefings with the state emergency manager and other department heads. We assessed the critical resources needed for each type of disaster—typical hurricanes in North Carolina. My job was to think ahead and recommend how active-duty soldiers, sailors, or airmen could be utilized as the disaster response progressed.

The effects of these disasters last longer than we realize. While a hurricane might pass through in a day or two, its devastation can take weeks, months, or even years to address. This prolonged response period quickly depletes the resources of National Guard soldiers and other volunteers with full-time jobs and lives outside disaster response. That's where the active duty military comes in.

Working alongside Colonel Jeff Maglio, the Emergency Planning Liaison Officer, and our NCO, a Master Sergeant, we maintained a 24/7 presence in the emergency operations center during disasters. Typically, Colonel Maglio would take the day shift, and I would cover nights unless circumstances required us to switch. Our role was to be constantly available to the governor and state emergency manager, ready to request active duty resources.

Then came 2020, and with it, a challenge none of us had anticipated: COVID-19. The impact of this virus was unprecedented, and immediately, all Emergency Planning Liaison Officers were mobilized for active duty. We reported to our Emergency Operations Center in Raleigh, offering the governor any necessary military assistance.

In those early days, ventilators and other medical equipment were critically short supply. The military stepped in to provide the resources we could to state and local governments. Our initial activation lasted about two weeks, and we thought our role might be over. But as the virus continued to spread and society as we knew it shut down, the NORTHCOM commander—the head of all U.S. military forces in North America—called us back to active duty to assist with the public vaccination effort.

Working with the Corps of Engineers, we identified 24 potential sites to serve as isolation hospitals across North Carolina, ranking them based on the time and resources needed to get them operational. Fortunately, we never exceeded our capacity, and this part of the plan wasn't activated.

Instead, we set up a vaccination center in an old Macy's store at a Greensboro, North Carolina mall. We created both an indoor vaccination facility, staffed primarily by airmen from Virginia Air Force bases, and drive-through lanes in the parking lot where people could receive their shots without leaving their vehicles. Throughout our operation, we administered more than 100,000 vaccinations.

We also coordinated with the National Guard to create roving teams of medics, providing mobile vaccination services in underserved areas for those unable to reach the main vaccination sites.

I served on active duty for over 90 days straight during this time, and I am proud to be part of such a crucial effort in combating the pandemic. It was pretty different from what I had imagined when I joined the National Guard in 2000. I never thought that being a part-time soldier would take me to war, through numerous natural disasters, or put me in charge of a COVID-19 vaccination center.

Looking back, I can see how my career came full circle. My love for law enforcement, emergency medicine, and emergency management wove through every stage, from volunteer firefighter and EMT to full-time fireman, police officer, and eventually, police chief.

Reflecting on my 24-year military career, I'm struck by how much it exceeded my initial expectations. From my early days as a young man eager to serve to my final roles in emergency management and counterterrorism, each experience built upon the last, shaping me into the person I am today.

I'm proud of our work to keep our country safe and assist in times of crisis. Whether it was responding to natural disasters, preparing for potential terrorist attacks, or helping to coordinate the response to a global pandemic, each challenge reinforced the importance of service, adaptability, and continuous learning.

My journey through the ranks of law enforcement, emergency services, and the military has taught me that you can make a significant impact no matter where you start with dedication and a willingness to learn and adapt.

Chapter 20

My Private Police Force

The unconscionable attacks on the United States on September 11th, 2001, didn't just change my trajectory as a future military officer - they reshaped my entire perspective on law enforcement. As the dust settled in New York, Washington D.C., and Pennsylvania, I knew the ripple effects would be far-reaching. We're taught that the goal of terrorism is to kill one and scare a thousand. On that day, they killed thousands and scared millions.

In the aftermath, I couldn't shake the feeling that our country's security landscape was about to shift dramatically. My studies and interest in domestic terrorism had prepared me to anticipate the public's reaction: people would be scared, and they'd want to see more uniforms, patrols, and visible signs of protection. But I also knew a harsh reality—there simply wouldn't be enough law enforcement officers to meet this surging demand.

I envisioned shopping centers without police presence and critical infrastructure left vulnerable. Armed security guards could fill some gaps, but I believed the situation required something more robust and professional. That's when an idea began to take shape, one that would eventually grow into a multimillion-dollar business and allow me to continue serving my community in ways I never imagined.

My mind drifted back to my experiences with private police departments in North Carolina. This concept had deep roots in the state's history. In 1872, not long after the Civil War, the governor signed the Company

Police Act into law. This innovative legislation allowed businesses or companies specializing in law enforcement services to operate under the governor's authority.

To truly understand the significance of this act, you have to picture North Carolina in those post-Civil War years. The state's economy revolved around cotton mills, tobacco farms, furniture makers, and clothing manufacturers. These were the lifeblood of many communities, often owning not just the factories but the stores, banks, and housing where their workers lived. The Company Police Act was a pragmatic solution to provide law and order in these company towns, especially in rural areas where sheriff's departments were stretched thin across vast counties.

Fast forward to the early 2000s, and this centuries-old concept suddenly felt incredibly relevant. I saw an opportunity to help supplement local police and sheriff's offices, filling crucial gaps in protection. My goal was clear - alleviate stress on traditional law enforcement by securing contracts with homeowners associations, shopping centers, business complexes, and others who felt they needed additional security beyond what their local agencies could provide.

With this vision in mind, I dove into starting a new business. The administrative hurdles were significant. I had to obtain all the requirements to form and operate a company police agency, also known as a special police agency. It was an intense 90-day sprint to secure the necessary approvals, complete testing, and navigate the complex paperwork.

But the real challenge emerged when I started looking for insurance. Finding a company willing to write a policy for an agency with the authority to arrest people, transport them, drive police cars with blue lights and sirens, and carry firearms? That proved to be a monumental task. This is a highly litigious area of work, and the likelihood of facing civil litigation was significant. It took another grueling ten months before I finally secured an insurance policy that met state requirements and gave me peace of mind that my assets wouldn't be on the line if something went wrong.

Finally, in August 2002, Capitol Special Police was born. The name was carefully chosen. In the early 2000s, we still relied heavily on phone books, so I wanted a name that would appear near the top of alphabetical listings. "Capital" also nodded to our base in Raleigh, North Carolina's capital city, where the local police department's patches read "Capital City Police." I liked the implication that we'd protect essential facilities, indicating our business's location.

Chief Taylor Police Car

Our initial team was small—just me and two other officers. We hit the ground running by going from one gated community to another, from apartment complexes to businesses, shopping centers, and wherever I thought there might be a need for extra police protection. Within weeks, we secured several full-time contracts and generated a steady income.

I knew our closest competitors would be local police departments offering off-duty officers for security work. To gain an edge, I set our hourly rates slightly below theirs. It was a delicate balance - we needed to be competitive and sustainable.

Those early days were a crash course in business management. Having spent most of my career in public service, with only a brief stint at Nortel Networks, I had a lot to learn about running a private company. Creating contracts, setting prices, advertising, and hiring employees - these skills I'd used as a police chief, but applying them in the private sector brought new challenges.

One of my first lessons came from a call with a prospective client - an apartment complex in Durham. The property manager's request caught me off guard: five police officers, seven days a week, 24 hours a day. Without thinking, I blurted out, "I don't think you can afford it." It was the wrong thing to say, and she quickly called me out. Backpedaling, I did some quick math and realized the cost would exceed $25,000 weekly. The manager agreed it was beyond their budget, but the exchange taught me the importance of listening first and offering solutions rather than making assumptions.

We scheduled an in-person meeting to assess their needs and budget constraints. What I discovered was sobering. The complex was grappling with entrenched gang activity, drug sales, and prostitution. Residents lived in fear due to frequent shootings. It was precisely the situation where our services could make a real difference.

I devised a strategy to deploy our officers during peak trouble times, ensuring at least one officer was on the property around the clock. Knowing that a strong initial presence would be crucial, I hired additional officers, most with years of law enforcement experience under their belts. We threw everything at the problem, often providing more manpower than we were being paid for.

The results were swift and dramatic. Within weeks, the shootings had diminished. The open-air drug markets vanished, and prostitutes no longer walked the streets openly. Slowly but surely, the community was returning to a safe place for residents.

Our success didn't go unnoticed. As the situation improved, the apartment community gradually reduced our hours, eventually scaling back to just one officer for 12 hours each night. It felt like a job well done - until the property was sold. The new owners refused to pay the outstanding bill for our last 60 days of service, leaving us with no legal recourse to recover the money.

This financial hit was a harsh lesson for my fledgling company. I still had to make payroll and pay taxes for the officers who had worked those shifts. The loss taught me the importance of billing in advance and being prepared to discontinue services when clients fall behind on payments.

Despite this setback, Capitol Special Police were off to a strong start. We surpassed $100,000 in revenue that first year, exceeding my initial expectations. When I spoke with our insurance agent about projected income, I modestly suggested we might hit $25,000. He laughed, insisting we'd clear $100,000 quickly. His confidence proved well-founded.

As we grew, I made sure our image matched our professionalism. Our patrol cars were fully equipped, mirroring city agency vehicles with blue lights, sirens, prisoner cages, and shotgun mounts. This visible deterrent was a crucial part of our strategy. We wanted potential criminals to know that the likelihood of getting caught was much higher on properties we patrolled.

Our officers were expected to handle the entire arrest process when necessary - from putting suspects in handcuffs to transporting them to the magistrate's office and then to jail. They gathered evidence and wrote reports that would be used in prosecutions. In essence, we operated as a fully functional private police force.

Our impact didn't go unnoticed by local law enforcement. I'll never forget hearing from a deputy chief at the Durham Police Department about a recent staff meeting. They reviewed city crime maps and noticed distinct "islands" with no criminal activity. When Police Chief Jose Lopez asked about these anomalies, one of his deputies said, "That's where Capitol Special Police is employed."

This validation from public law enforcement was immensely gratifying. We were proving ourselves as true force multipliers, reducing calls for service in our contracted areas and allowing city officers to be more proactive elsewhere.

One of our most significant contracts came from a local shopping mall desperate for help with traffic control during the holiday shopping season. Customers were facing 45-minute waits to exit the parking lot. Gridlock was spilling onto public roadways, and fender-benders were common.

This presented a unique challenge. Company police officers don't have authority on public roadways, but I found a creative solution. North Carolina motor vehicle law allows police chiefs or sheriffs to appoint traffic control officers - typically used for firefighters or ambulance attendants. I approached Sheriff Donnie Harrison, whom I knew from my days as a deputy sheriff, and asked if he'd designate my officers for this purpose. He agreed without hesitation.

We also offered to handle motor vehicle crashes in the mall parking lot, relieving Raleigh PD of that responsibility. Most officers dread working these minor accidents - they're typically low-damage incidents caused by momentary inattention but still require the same paperwork as a major highway collision. By taking this off their plate, we cemented our reputation as valuable partners in the law enforcement community.

As Capitol Special Police grew, I faced a new challenge—keeping up with hiring. Getting new officers approved by the state took at least 90 days, slowing our ability to meet contractual needs. I obtained a security guard license from the state's Private Protective Services Board to address this. This required finding someone with extensive security management experience to serve as the licensee, as my status as a law enforcement officer disqualified me.

I turned to my network at Nortel and recruited their director of security, Jim Gillespie. Jim became integral to our new business arm, Capitol Special Patrol. He quickly helped assemble a team of professional security officers who could assist our police officers on multi-officer contracts or take on pure security assignments.

Our expansion continued when a business acquaintance in Charlotte approached me about selling his security company. It was an exciting opportunity, but it required financing to purchase. With help from bankers at SunTrust, I secured a small business administration loan. This acquisition doubled the size of the Capitol Special Patrol overnight.

Moving into the Charlotte market was a two-and-a-half-hour drive away, which brought new challenges. I needed to find trustworthy managers who shared my ideals and work ethic. Fortunately, I connected with experienced supervisors who helped our new office location thrive.

The growth didn't stop there. Years later, I had the opportunity to purchase Blue Ridge Public Safety and Sapphire Valley Public Safety. These businesses served the picturesque resort communities in the southern part of Jackson

County, North Carolina. It was an area that swelled from 2,800 year-round residents to over 40,000 during peak vacation times.

This unique environment was perfect for a company police agency. Many property owners were out-of-state residents who couldn't vote in local elections despite paying taxes on multi-million dollar estates. This led to minimal public resources for law enforcement, creating a niche that private security could fill.

Acquiring these mountain agencies in 2014 marked a turning point. I finally left my full-time position with the Virginia National Guard to focus entirely on running my five businesses. It was a leap of faith backed by years of hard work and careful planning.

Throughout this journey, I've held fast to a core philosophy - treating people with empathy and respect, even in difficult situations. I ask my officers to approach every interaction as if dealing with their family members. How would you want an officer to treat your mother if she was intoxicated and causing a disturbance? This perspective helps de-escalate tense situations and often leads to better outcomes for everyone involved.

Building these businesses hasn't been without its challenges. I've had sleepless nights worrying about making payroll or finding funds for essential equipment. I vividly remember Christmas 2004, when no client had paid their November invoice. My family was scheduled for a Bahamas cruise; tickets were purchased months in advance. I had to make the gut-wrenching decision to delay employee paychecks, explaining the situation to each one personally. It was a hard lesson in cash flow management and the importance of building a financial cushion.

Despite the occasional setbacks, the rewards of this work are immeasurable. I take immense pride in introducing myself as the police chief for Capitol Special Police or Blue Ridge Public Safety. Knowing that we're making communities safer, giving people peace of mind, and complementing the work of public law enforcement - it's deeply fulfilling.

If you dream of starting your own business, especially in a field you're passionate about, I advise you to do your research and reach out to others in the industry, even if they're not your direct competitors. When I was starting out, I connected with Dave Finn, the previous owner of Blue Ridge Public Safety and Sapphire Valley Public Safety. His insights were invaluable in helping me navigate potential pitfalls.

As you grow and achieve success, you must pay it forward. I've made it a point to mentor others looking to enter this field by sharing resources and knowledge to smooth their path. In this industry, as in life, we rise by lifting others.

I firmly believe that authentic leadership involves delivering exceptional services to your clients and cultivating the next generation of professionals. By providing guidance and mentorship, you empower your team to uphold the highest service standards and contribute to the safety and well-being of the community you serve. Through mentorship, you strive to pass on the knowledge and expertise you have gained throughout your career, fostering a culture of continuous learning and excellence within your organization.

As I look to the future, I see continued growth for private law enforcement and security services. The need for protection and peace of mind isn't going away. It's a responsibility I take seriously and continues to drive me forward. Although every day may bring new challenges, it also brings new opportunities to serve and make a difference. And really, isn't that what a daily life of service is all about? So, go on and make a difference in your line of service.

Chapter 21

Federal Court Expert Witness

In October 2013, a seemingly routine firearms training session set me on an unexpected path that would redefine my career and allow me to continue serving justice in ways I never imagined. Mr. Mel Tucker, a distinguished retired police chief from Tallahassee, Florida, had scheduled a course with me to renew his HR 218 carry certificate. Little did I know that this encounter would open up a whole new world of opportunity.

HR 218, also known as the Law Enforcement Officer Safety Act, was signed into law by President Bush in 2004. This legislation allows retired law enforcement officers and select others who carried firearms during their careers to continue doing so in retirement. The idea was to have more trained individuals on the streets, especially during heightened terrorist activity. To maintain this privilege, officers must complete an annual state-mandated firearms qualification.

The course I conducted for Mr. Tucker was comprehensive. We covered day and night fire scenarios, ranging from close-quarters combat at three yards to precision shooting at 25 yards. In the classroom, we delved into firearms maintenance, honed marksmanship skills, and reviewed the intricate laws governing the use of deadly force at both federal and state levels.

As we wrapped up the training, Mr. Tucker turned to me with an appraising look. "Wow, you're good at this," he said, his voice tinged with admiration. "Have you ever thought about being an expert witness?"

I was taken aback. Me? An expert? The thought had never crossed my mind. "No," I replied, shaking my head. "I don't think I'm an expert. I have a friend, Doug Scott, a retired police officer specializing in drug recognition. He often testifies in driving while impaired cases, but that's not me."

Mel's response was immediate and emphatic. "You are an expert," he insisted. "If you're interested in learning how to do this, let me know. I'll gladly teach you, give examples of reports I've written, and even help you get your first case."

His confidence in me was both flattering and intriguing. Without hesitation, I agreed, and we set a date to meet at his home. When I arrived, Mel greeted me warmly and ushered me into his study. There, he presented me with three meticulously written reports on cases involving police officers. As he walked me through each one, I began to see the depth of analysis and expertise required in this field.

Mel also gave me a list of references he no longer intended to use. He was hoping to reduce his caseload and ease further into retirement. His generosity and willingness to mentor me were truly remarkable.

True to his word, a few weeks later, Mel connected me with an attorney representing two police officers from Tallahassee. This would be my first case as an expert witness, and I was equally excited and nervous.

The case revolved around a tragic incident at a fraternity initiation party for rising high school seniors. The local Lions Club had contracted two off-duty police officers to provide security for the event. The goal was to expose teenagers to fraternity life and encourage them to enroll in college and join their organizations.

I could almost see the scene unfolding as I delved into the details. The party was in full swing, music pulsing through the air, young faces alight with excitement about their future. But beneath the surface, tension was brewing.

A local gang member had infiltrated the group and started causing a disruption. The two officers swiftly escorted him outside and made him leave the property. They then returned to monitor the event from inside.

However, the atmosphere had shifted. The teenagers became increasingly rowdy, and the Lions Club members, growing concerned, decided to end the party an hour earlier than planned. The officers began ushering the youths into the parking lot, where many would have to wait for rides home.

What happened next was a chilling reminder of how quickly a situation can spiral out of control. The ejected gang member, seething with resentment, had retrieved a firearm and was lying in wait in the wooded area surrounding the parking lot.

Suddenly, a shot rang out, cutting through the night air. A promising 18-year-old student fell to the ground and was shot in the head. He died at the scene.

Chaos ensued. I could imagine the officers' adrenaline spiking as they tried to determine where the shot came from, calling for backup and working to preserve evidence for the detectives who would soon arrive.

In the aftermath of this tragedy, the victim's family sought justice through legal channels. They filed a lawsuit against the Lions Club and the two off-duty officers, alleging negligence in their security duties.

This is where I came in. As an expert witness, I explained to the jury how these officers should have conducted themselves at this event. I needed to address the plaintiff's attorney's allegation that the officers should have checked the wooded area before allowing students into the parking lot.

As I prepared my analysis, I drew on my years of experience in law enforcement. I knew that the likelihood of a sniper lying in wait after a teenage party would be statistically insignificant. Without specific information suggesting otherwise, officers wouldn't typically think to search the woods for someone they had previously ejected from the property.

In the world of expert witnesses, there are two main categories: scientific and non-scientific. As a police procedure expert, I fall into the latter category. My expertise is based on my experience, training, and education in law enforcement. To be accepted by the court, I have to demonstrate that I have sufficient specific knowledge in the area in question to offer an informed opinion to help the jury decide the case.

The central question was: Were the police officers responsible, or did they contribute to this young man's death? Or did they fulfill their responsibilities based on nationally accepted police procedures and practices?

After a thorough analysis, I concluded that the officers had indeed done everything that would have been expected of them under the circumstances. Given their information at the time, the officers could not have reasonably foreseen or prevented the tragic situation.

Ultimately, the Lions Club and the city offered a settlement to the young man's estate, and the lawsuit was dropped. While the outcome was bittersweet—no legal decision could truly compensate for the loss of a young life—the case gave me invaluable experience in drafting my first expert report, finding literature to support my opinions, and navigating the complex world of legal testimony.

As I continued to take on more cases, I learned that being an expert witness often involves being deposed. This is when the opposing side's attorneys question you about your report, how you derived your opinions, and how you're supporting those opinions. Their goal is often to disqualify you and prevent your opinions from being presented in court.

These depositions can be grueling experiences. The attorneys might try to upset you or catch you off guard. They're looking for any weakness that could show you don't have the necessary education, training, or experience to be accepted as an expert witness. It requires a thick skin and unwavering confidence in your expertise.

I'm proud to say I've never been excluded as an expert witness. When I stand to testify, I introduce myself as "Dr. Colonel Chief Roy Taylor." It might sound pretentious, but there's a method to it. I want the jury to understand immediately that I have a Ph.D. in criminal justice, served 24 years in the United States Army Military Police Corps, and am an active chief of police with over 30 years of experience in federal, state, local, and private law enforcement agencies.

My approach to each case is methodical. I review it through the lens of a police chief, imagining a citizen coming into my office with a complaint about my officers violating their civil rights. I ask myself: Did my officers follow our agency's policies and procedures? Are our procedures in line with nationally recognized standards? Did my officers obey the law?

Most of my cases fall under what's known as "1983 cases," referring to Title 42, Section 1983 of federal law. This statute prohibits law enforcement or other government workers from knowingly violating someone's civil rights. Most often, these involve Fourth Amendment rights and allegations of illegal seizure, though I've also worked on cases involving First, Fifth, Sixth, and Fourteenth Amendment rights.

Over the years, I've handled over 225 cases, most of which were federal 1983 cases about police procedures. Many involve shootings, including what is termed "negligent discharges" - when an officer fires their weapon accidentally.

I've found myself educating jurors on the four cardinal rules of gun safety, emphasizing that officers are trained never to put their finger on the trigger until they are ready to fire and have the legal authority to do so. Unfortunately, fear often leads officers to keep their finger on the trigger, resulting in accidental discharges through what's known as a sympathetic response.

Another common type of case I handle involves high-speed motor vehicle pursuits. These cases often result in death or serious injuries, not just to the suspects or officers involved but to innocent bystanders as well. I've seen a significant shift in police policies over the years, with many agencies now restricting pursuits to only the most serious violent felonies.

One particularly heartbreaking case involved a 23-mile pursuit through the winding roads of West Virginia. The initial infraction? A driver forgot to turn on their headlights when pulling out of a parking lot. The pursuit ended with a rollover crash that left a young mother, a quadriplegic, never able to hold her baby again.

Cases like these underscore the critical importance of sound judgment and continuous reevaluation of circumstances during pursuits. Officers need to weigh the potential risks to society against the severity of the initial offense. Is apprehending someone who forgot to turn on their headlights worth endangering lives? In my expert opinion, absolutely not.

Being an expert witness isn't always popular within law enforcement circles. Some officers view us as "wolves in sheep's clothing" or liken us to internal affairs divisions, often derogatorily called the "rat squad." But I firmly believe our role is crucial in ensuring justice prevails and people's rights are protected. After all, isn't that the oath we take when we become police officers?

I'll always be grateful to Mel Tucker for recognizing my potential and mentoring me in this field. His guidance has allowed me to continue serving justice long after my active duty days, impacting current and future law enforcement operations to ensure they stay within the bounds set by the United States Constitution.

Law enforcement officers need to be well-trained and continuously updated on police procedures and tactics to ensure they don't exceed what is legally permitted. It's often portrayed on TV that police officers or detectives will use unauthorized means to capture the "bad guy" before the show's end. This is not reality but entertainment. The fact is that most police procedures are governed by civil rights and case law derived from previous litigation that has resulted in specific restrictions on law enforcement operations.

Throughout my career as an expert witness, I've seen the profound impact of Supreme Court decisions on law enforcement practices. One such case was Tennessee v. Garner, which I believe was decided in 1984. It involved a young man who had broken into a house and fled from the police. The officers shot and killed him over stealing $20 from an unoccupied house. This case significantly changed how law enforcement could use deadly force against fleeing suspects.

Another landmark case came in 1987 when the Supreme Court handed down Graham v. Connor. This case involved excessive force by the Charlotte, North Carolina police department. The ruling established the "objective reasonableness" standard for evaluating the use of force by police officers. It allows officers to use force whenever they believe the person is acting suspiciously, attempting to flee, or making threats. The decision permits officers to overcome that resistance with whatever level of force is warranted, but it must be objectively reasonable given the circumstances.

We've also seen the case of Arizona v. Gant, which restricts when officers can search motor vehicles they have stopped. If the person is handcuffed and no longer near the vehicle, police officers must secure a search warrant or obtain consent from the driver to search the vehicle unless they can justify some exigent circumstance.

These cases typically result from law enforcement officers overstepping the bounds set by the Constitution, requiring the court to define those restrictions further. While these rulings often limit law enforcement powers, the justices aim to ensure that citizens are treated fairly and their constitutional rights are protected.

I take my role as an expert witness very seriously and appreciate the opportunity to continue serving the community in this capacity. I will always be grateful to Mr. Tucker for showing me this other law enforcement-related career path. It's opened up a world of possibilities for me to contribute to law enforcement and justice.

It's worth noting that expert witnesses are used in almost every facet of any profession. Dentists, doctors, nurses, engineers - anything that can result in litigation probably has someone designated as an expert witness to assist the judge and jury in determining culpability based on these experts' testimony. We can't know the intricacies of every profession, which is why expert witnesses play such a crucial role in our legal system.

Even if you're not a law enforcement officer, there may be opportunities for you to become an expert witness once you have garnered sufficient experience, education, and training in your field. You might be in a scientific field and be deemed an expert based on your ability to deal with chemicals or compounds that result in an injury or death. I encourage you to look into these opportunities if you are a senior in your career field. It may open up a world of new possibilities for you, as it did for me.

Being an expert witness has allowed me to continue serving my community and the cause of justice long after my active duty days. It's a role that requires constant learning, critical thinking, and a deep commitment to fairness and truth. Every case I take on is an opportunity to ensure that the law is applied correctly and that officers and citizens are held to our constitution's standards.

Reflecting on this chapter of my career, I'm surprised by how it encapsulates so much of what I've learned throughout my service life. The attention to detail I honed as a Boy Scout, the discipline I learned in the military, the investigative skills I developed as a police officer, and the leadership experience I gained as a chief come into play in my role as an expert witness.

This work has also given me a unique perspective on law enforcement's challenges today. I've seen firsthand how policies and procedures can make the difference between life and death, justice served, and rights violated. It reinforced my belief in the importance of continuous training and education for law enforcement officers and the need for clear, well-thought-out policies that balance public safety with individual rights.

To those of you reading this who may be considering a career in law enforcement or who are already serving, I urge you to always remember the gravity of the public's trust in you. Every decision you make, and every action you take has the potential to impact lives profoundly. Strive to always act with integrity, continuously educate yourself, and uphold the highest standards of your profession.

For those of you in other fields, I encourage you to consider how your expertise might be used to serve justice. Whether as an expert witness or in some other capacity, there are always opportunities to use your knowledge and experience for the greater good.

As I continue in this role, I remain committed to the principles that have guided me throughout my career: service, integrity, and the pursuit of justice. It's a challenging path but one that I find deeply rewarding. I believe this path is vital to ensuring that our justice system remains fair, balanced, and true to the ideals upon which it was founded.

In closing this chapter, I want to emphasize that being an expert witness is more than just a job or a role—it's a responsibility to the truth, justice, and the continuous improvement of law enforcement practices. It's a way to ensure that the lessons learned over a long career in service can continue to make a positive impact long after the uniform has been hung up.

So, whether you're just starting in your career or a seasoned professional looking for new ways to make a difference, I encourage you to consider how your expertise might be used to serve others. You never know where it might lead you - just as I never imagined, back in that firearms training session with Mel Tucker, that I was about to embark on an entirely new chapter of my life of service.

Chapter 22

Becoming a Doctor

After three years as an expert witness, I realized that pursuing a Ph.D. in criminal justice would enhance my expertise and solidify my standing in the courtroom. I'd already been deposed several times and testified in a couple of trials, and I knew that to excel in this field truly, I needed impeccable credentials. In the world of law enforcement and expert testimony, credibility is everything.

My journey to becoming Dr. Taylor began with a simple yet crucial decision: choosing the right university. I needed an institution with a stellar reputation and, most importantly, proper accreditation. After thorough research, I set my sights on Walden University. With over 60,000 students and accreditation in Minnesota, Walden offered precisely what I was looking for - a robust criminal justice degree program.

The application process involved no walking in the park. I had to submit transcripts from previous colleges, complete a writing test, and explain my motivations for seeking an advanced degree. There were also in-depth conversations with academic advisors. But I was determined, and my efforts paid off when I was admitted into the Ph.D. program.

There was a catch, though. I would have to simultaneously complete a master's degree in criminal justice alongside my Ph.D. studies. This meant a heavier course load, but I've never been one to shy away from a challenge. So, I dove into my studies, driven by the knowledge that this additional qualification would only strengthen my expertise.

By 2016, I had completed all the requirements for both the master's and Ph.D. programs, save for one final keystone project - the dissertation. I was told that the average student takes five years to complete a Ph.D., with some taking up to eight years. But I'm not your average student. My flexible work schedule and unwavering work ethic determined me to beat that average.

Choosing a dissertation topic was quite complex. If it is too broad, your research will lose significance. It might not be comprehensive enough to warrant a doctorate if it is too narrow. As I pondered my options, I considered various areas of the criminal justice system that intrigued me. Community policing has always been a passion of mine, particularly its potential to bridge gaps in communities that often feel over-policed.

But then, in November 2016, something extraordinary happened. In just one week, I received six cases involving officer-involved shootings of individuals in psychiatric crisis. This wasn't just a coincidence but a clear sign of a systemic issue that demanded attention. At that moment, I knew I had found my dissertation topic.

I decided to focus on the Crisis Intervention Team (CIT) program, which originated in Memphis, Tennessee, in 1988 following a series of police shootings involving individuals with mental health issues. As I delved into existing research, I noticed a significant gap. Most studies were quantitative, focusing on statistical effectiveness. What was missing was a qualitative approach - one that explored the human experiences and perceptions of the officers on the front lines.

The CIT program isn't just about training police officers. It's a collaborative effort involving mental health care workers, social workers, emergency room doctors, and law enforcement. Together, they form a coalition to divert people suffering from mental health crises away from the criminal justice system and toward the help they truly need.

This approach is crucial, given the stark reality that our local jails, state prisons, and federal correctional facilities have essentially become the most prominent hospitals for people with mental illness. The majority of inmates in county jails struggle with some form of mental illness or substance abuse issue, requiring a small army of mental health workers, psychologists, and psychiatrists to address their needs.

To truly understand the gravity of this issue, we need to step back in time to 1963. President John F. Kennedy, whose own sister had undergone a frontal lobotomy to treat her mental illness, was advised that individuals like her would be better served by local mental health resources rather than state psychiatric hospitals. The idea was to improve their quality of life by bringing treatment closer to home.

However, the execution of this plan was tragically flawed. Funding for state mental health facilities was withdrawn before local resources were established. When President Kennedy was assassinated shortly after, the initiative lost its champion. President Johnson attempted to continue the work but lacked the political capital to make it happen. The result was that thousands of people with mental illness were suddenly without care, marking the beginning of our modern homeless population.

This shift inadvertently thrust law enforcement officers into the role of de facto mental health crisis responders. The problem is that most officers lack adequate training to deal with people in crisis. They often fall back on their academy training, focusing on command presence and the "ask, tell, and make" approach. This method, while effective in many law enforcement scenarios, can be disastrous when dealing with individuals in mental health crises.

People experiencing a mental health crisis often perceive reality differently. They may be hearing voices, experiencing delusions, or struggling with medication issues. The typical law enforcement approach of commanding compliance can escalate the situation, sometimes with tragic consequences.

This is where the CIT program comes in. It provides 40 hours of specialized training to volunteer officers, teaching them about different types of psychosis, medications, and de-escalation techniques. They learn about community resources and alternatives to incarceration. The program emphasizes empathy, active listening, and building connections before attempting to direct or control the situation.

Considering this background, I conducted a qualitative study focusing on CIT training. I wanted to interview officers currently on crisis intervention teams, exploring their experiences in the field and their personal beliefs about which aspects of the training were most effective.

The process of writing a dissertation is rigorous and structured. I selected my chair and committee member, both of whom were intrigued by my focus on mental health issues in law enforcement. With their approval, I began the intensive process of reviewing the literature. I read over 200 empirical articles about CIT training, meticulously organizing my notes and citations in an Excel spreadsheet.

Creating the survey instrument was a crucial step. I had to carefully craft questions that would elicit my needed responses while ensuring they were open-ended enough to generate meaningful data. Once my committee approved my literature review and survey, I submitted them to the Institutional Review Board (IRB) to ensure my study wouldn't harm participants.

Then came an unexpected challenge - the COVID-19 pandemic. The world went into lockdown just as I was ready to begin my interviews. However, I adapted and switched to video conferencing platforms like Zoom and Teams. This unexpected change proved beneficial, allowing me to record interviews and focus more on facial expressions and follow-up questions.

I interviewed two active CIT officers from each participating Wake County, North Carolina municipality. Eleven out of thirteen agencies agreed to participate, giving me a robust sample size. The officers were enthusiastic about the study and eager to share their perspectives on the CIT program. They unanimously supported the program and called for regular refresher training to stay current on techniques and treatments.

Analyzing the data was also a monumental task, but technology again came to my rescue. I used specialized software to identify patterns and frequencies in the officers' responses, making the analysis process more manageable. This allowed me to draw meaningful conclusions and complete my dissertation.

The day I presented my findings to my committee members was one I'll never forget. After four years of intensive study and work, hearing "Congratulations, Dr. Taylor" was an indescribable feeling of accomplishment.

Completing a Ph.D. while managing two law enforcement agencies and three security companies and serving as a lieutenant colonel in the Army Reserve was no small feat. It required countless 20-hour days, unwavering dedication, and the ability to juggle multiple responsibilities. It's a journey that tested my limits, expanded my horizons, and ultimately transformed me personally and professionally.

Let me tell you, the path to earning those three letters after your name is not for the faint of heart. I've met countless individuals who began their PhD journey only to abandon it, overwhelmed by the sheer time and effort required to complete the dissertation. It's a grueling process that demands unwavering dedication and sacrifice. But for those who persist, the rewards are immeasurable.

Now, when I walk into a courtroom as an expert witness, the title "Doctor" carries significant weight. It's not about ego or prestige – it's about credibility. That title immediately signals to the jurors that I'm not just a seasoned law enforcement officer but also someone with the academic credentials to back up my expertise. It tells them I can conduct rigorous research, analyze complex data, and form well-supported opinions.

This academic background has been invaluable in my work as an expert witness. When formulating opinions for a case, I can dive into published studies and scholarly works to bolster my arguments. It's given me a new toolkit for approaching complex law enforcement and criminal justice issues.

You might be surprised to learn that only about 1% of the world's population holds a PhD. We're inching closer to 2% in the United States, but it's still a rare achievement. I believe this speaks to both the challenge of earning a doctorate and the growing recognition of its value in advancing one's career and contributing to society.

I often joke that I might become a "2 %-er" in this role. But in all seriousness, I firmly believe that the more educated our society becomes, the better off we'll all be—primarily when those with advanced degrees actively engage in their communities and apply their knowledge to solve real-world problems.

To anyone considering this monumental task, I say go for it. Yes, it's difficult. Yes, it's time-consuming. And yes, you'll have to prioritize it in your life. But the personal and professional growth you'll experience is unparalleled.

My journey as a lifelong learner has developed me in ways I never anticipated. Beyond accumulating knowledge, pursuing advanced education has made me a more critical thinker, problem-solver, and empathetic leader. The rigors of academic study sharpened my analytical skills, but perhaps more importantly, it taught me to approach challenges from multiple perspectives.

This is invaluable in law enforcement, especially in leadership roles. There's rarely a single, clear-cut solution to our daily complex problems. Balancing public safety, justice, and the individual rights of citizens, particularly those struggling with mental health, requires nuanced thinking. It's no longer enough to apply the same old methods—we must innovate, adapt, and stay open to new ideas.

My Ph.D. journey also deepened my appreciation for collaboration. Throughout the dissertation, I had to rely on others - my academic advisors, the officers who participated in the study, and even the mental health professionals who helped me understand the subject matter more deeply. Their insights, perspectives, and critiques improved me, just as collaboration in law enforcement leads to stronger, more effective outcomes.

This doctoral experience enhanced my professional capabilities and transformed me personally. There were moments of doubt when I questioned if I could balance everything on my plate. But I pushed through those doubts because I knew what was at stake. My commitment wasn't just to the degree itself – to the principles I believed in, to improving the criminal justice system, and to being the best version of myself.

Achieving a Ph.D. has allowed me to merge my real-world experience with academic theory, creating a powerful synergy in my roles as an expert witness, a law enforcement leader, and a public servant. As I said earlier, the title of "Doctor" isn't just about prestige; it represents a deep commitment to pursuing knowledge and applying it to solve practical problems.

That, ultimately, is why I encourage anyone who feels the call to pursue advanced education. It's not just about the degree—the journey of self-discovery, the intellectual rigor, and the contribution you can make to your field and society. Higher learning equips you to think critically, act decisively, and lead purposefully, whether in criminal justice, healthcare, education, or any other discipline.

My educational journey was far from linear. It took me eight years to earn my two-year associate's degree, juggling classes around my rotating shifts as a police officer. But I persevered, and that accomplishment laid the foundation for everything.

Next came my bachelor's degree from Mount Olive College, which I completed in two years. Later, while working as an emergency manager in the civilian sector, I earned my first master's degree in occupational safety from East Carolina University. This expanded my expertise beyond my core passion for law enforcement, giving me a more well-rounded skill set.

But I didn't stop there. I added a master's degree in criminal justice to my resume, followed by the crowning achievement - my PhD in criminal justice. Each step of this journey required dedication, sacrifice, and a willingness to prioritize education over other aspects of my life.

Looking back, I can confidently say: if I can do it, you can, too. The key is to dedicate yourself fully to the process. Be prepared to give up some hobbies and leisure time. Understand that this continuous development will open doors and create opportunities you never thought possible.

Becoming Dr. Taylor was never the end goal—it was the beginning of a new chapter. This chapter continues to unfold as I use the tools and knowledge I've gained to make meaningful contributions to law enforcement and beyond. Whether it's through mentoring the next generation of officers, shaping policies prioritizing mental health, or standing in front of a jury as an expert witness, I carry the lessons learned along this remarkable journey.

As I reflect on where this path has taken me, I'm filled with gratitude for the title and the experience that came with it. It's a reminder that the most significant accomplishments in life aren't just about reaching a destination; they're about the person you become along the way. And for that reason, the title of "Doctor" will always hold a special place in my heart, not just for what it represents but for how it has shaped my character and my contributions to the world around me.

Daughter and Grandsons

Chapter 23

Mental Health and Law Enforcement

- From Academia to the Streets -

As we transition from the world of academia and my pursuit of a Ph.D., I want to underscore how my educational journey directly influenced my approach to one of the most pressing issues facing law enforcement today: mental health crisis response. The analytical skills and research methodologies I honed during my doctoral studies proved invaluable as I delved deeper into this critical area.

Over my 45-year career in law enforcement, as well as my time as a firefighter and emergency medical technician, I've had countless encounters with individuals exhibiting signs of mental illness. But it wasn't until I began my work as an expert witness on police procedures, armed with the academic rigor from my Ph.D. studies, that I truly grasped the magnitude of this issue.

I'll never forget that week in November 2016 that involved six cases of officers shooting people in a psychiatric crisis. The sheer volume was staggering, and it lit a fire in me. Drawing on my research skills in law enforcement and academia, I knew I had to dig deeper to understand this problem and find a way to help law enforcement more effectively deal with people suffering from mental illness.

My goal became twofold: reduce the number of fatalities in these encounters and help officers understand that there are alternatives to funneling these individuals into the criminal justice system. It's a complex issue, but I often frame it this way: Would we treat people suffering from cancer or heart conditions in the same manner? Of course not. Yet, people with mental illness often face a stigma that leads to unconscious biases, even among well-meaning officers.

Let's break it down by applying the analytical approach I cultivated during my Ph.D. studies. When law enforcement responds to calls involving emotionally disturbed individuals, they're not just acting as police officers. They suddenly find themselves thrust into the roles of mental health worker, social worker, and community support service – often without proper training or clear directives on how to navigate these complex situations.

The numbers are sobering. According to a study by the Treatment Advocacy Center, over 8 million Americans have a severe mental illness that impairs their thought processes. Almost half of these individuals don't receive treatment, primarily due to the deinstitutionalization of state-run mental health facilities and other policy failures.

This untreated population accounts for 10 to 20% of all calls that law enforcement officers respond to. Too often, the outcome is arrest and incarceration rather than a connection to mental health resources. For juveniles, it might mean placement in a detention center instead of getting the help they desperately need. These initial encounters often trap emotionally disturbed individuals in a vicious cycle of arrest, incarceration, and recidivism.

But it gets worse. These encounters frequently require officers to make split-second decisions about a person's intentions, often resulting in the use of force. The Treatment Advocacy Center found that people with severe mental illness were 16 times more likely than the general population to be killed by a law enforcement officer. That represents nearly 25% of all fatal police encounters. Let that sink in for a moment.

This stark reality, coupled with my experience as both a police officer and expert witness, made it crystal clear: we need more training, particularly scenario-based training, so officers can learn firsthand how it feels to interact with someone in a psychiatric crisis.

One answer to this challenge emerged from tragedy. In 1987, Memphis, Tennessee, witnessed the shooting death of an African-American man experiencing a mental health crisis. It was the third such incident in Memphis that year, and the community had had enough. They expressed their outrage peacefully, convincing Mayor Wharton to establish a working group. This coalition brought together law enforcement agencies, mental health professionals, and advocates from NAMI (the National Alliance on Mental Illness).

Their mission was to develop a program to reduce the risk of injury to both officers and emotionally disturbed individuals while diverting people from the criminal justice system and increasing access to mental health treatment. Their efforts resulted from the Crisis Intervention Team (CIT) training program, now known as the Memphis Model.

By 2016, when I began researching for my dissertation, over 2,800 police departments across the United States had implemented this 40-hour training program. It's even been replicated in Australia, Canada, and the United Kingdom, underscoring that this isn't just an American problem – it's a global challenge.

I've thrown myself into this work. I've completed the initial CIT training and the CIT coordinator training, which equips me to facilitate collaborative agreements between local police departments, mental health agencies, emergency rooms, and treatment facilities. I've also become certified as a

CIT trainer, allowing me to instruct officers and emergency workers on the program's core principles. Additionally, I serve on the state steering committee for the CIT program, providing input on program development and funding.

When something significant calls you to assist, you must do whatever possible to answer that call and make society better. This issue has caught the attention of government leaders at all levels, primarily due to the intense scrutiny law enforcement has faced in recent years over excessive force situations. While there's widespread agreement that current funding is insufficient to provide the necessary resources to address this mental health crisis effectively, progress is being made.

We're also grappling with the growing opioid epidemic and the high rates of post-traumatic stress disorder among returning war veterans. The increasing homeless population has further strained available mental health resources in local communities. Ultimately, these issues fall on the shoulders of law enforcement, expanding their role to include that of emergency mental health worker - a position that requires specialized training.

In North Carolina, we've increased our basic law enforcement training curriculum to provide new officers with 24 hours of instruction on recognizing and dealing with people in mental health crises. The International Association of Chiefs of Police has developed an eight-hour training program titled "Mental Health First Aid," another step towards educating officers about recognizing and responding to psychiatric crises.

The Police Executive Research Forum has also created an excellent 40-hour ICAT program (Integrating Communication Analysis Tactics). This curriculum teaches officers techniques and procedures for responding to crises by employing distance and shielding, affording more time to de-escalate the situation, and hopefully connecting individuals with resources outside the criminal justice system.

Let me share a case that illustrates the critical importance of this training. I was retained as an expert in a situation involving a man in Kentucky diagnosed with schizophrenia. One afternoon, feeling the onset of delusions, he agreed to his fiancée's suggestion to admit himself to the local hospital for treatment. On the way there, they passed a Kentucky state trooper conducting a traffic stop, lights flashing. This triggered a paranoid delusion in the man, convincing him his fiancée was leading him into a trap.

Despite her attempts to reassure him, he became increasingly agitated. She exited the highway, hoping to take back roads to prove she wasn't deceiving him. When they stopped at the end of the exit ramp, he seized the opportunity to flee, pulling out a small pocket knife as he ran into a nearby subdivision.

His fiancée called 911, explaining the situation and requesting that police not respond, fearing it would escalate his paranoia. The dispatcher informed two patrol units, but they responded to the area anyway. The first officer met the fiancée outside the subdivision and agreed to stay back. However, the backup officer didn't stop to learn more and immediately entered the area, quickly locating the man.

Both officers eventually confronted the man standing in a vacant field. They called him to approach, guns drawn but partially concealed. His fiancée ran to him, hugging him and begging him to put away the knife. The officers yelled at her to get away, and when she finally did, they ordered the man to drop the knife. When he didn't comply immediately, they opened fire, shooting him multiple times.

The officers claimed he was moving toward them and within 21 feet - a distance often cited in police training as a danger zone for knife attacks. However, the fiancée insisted he was turning to run. The autopsy revealed bullet wounds in his side and back, lending credence to her account.

Both officers had received crisis response training - one was a certified CIT officer, the other a hostage negotiator. In my expert opinion, this tragedy could have been avoided if the officers had positioned themselves on the opposite side of their patrol cars, creating a barrier between themselves and the distressed individual. This would have given them more time and options to de-escalate the situation.

The police chief of this agency had publicly stated a goal of having 100% of officers trained in CIT concepts. While I agree that all public safety personnel should have some level of mental health crisis training, NAMI recommends that only officers who volunteer and are selected undergo the entire CIT program. This is because these individuals often exhibit more empathy for affected populations.

This case highlights that merely going through training doesn't guarantee its principles will be implemented effectively. The officers could have allowed more time for the fiancée to de-escalate the situation or let the man leave to continue his journey to the hospital. At that point, he hadn't committed any crime, and the officers' presence only escalated the situation, just as his fiancée had feared.

I've encountered many similar cases over the years, all pointing to the same conclusion: police officers need more specialized training to deal with these situations. They must learn to differentiate between responding to a call involving a fleeing felon - like a bank robber or murderer who has intentionally committed a crime - and a person in a psychiatric crisis who isn't entirely grounded in reality.

These individuals are experiencing an alternate universe that we can't understand. But with patience and time, we may be able to make a connection that allows us to help them receive treatment from local resources. That's the ultimate goal of the Crisis Intervention Team - to affect these situations positively instead of treating them as criminal matters that too often end in the use of deadly force.

Another profoundly troubling scenario I've encountered involves police officers shooting and killing individuals attempting suicide. It's worth noting that in the United States, only one state has a law against suicide. So, in 49 states, it's not illegal. If someone is threatening to harm themselves but not posing an immediate threat to officers or others, what justification is there for using deadly force?

The prevalence of civil litigation against law enforcement agencies in these cases has led some departments, primarily in the western United States, to stop responding to calls of suicide in progress altogether. They inform callers that it's not a police matter but a mental health issue, and they need to contact someone else. While this approach may seem harsh, it's somewhat understandable given the large settlements awarded to plaintiffs' estates in cases where law enforcement failed to prevent suicide or, in some instances, were deemed responsible for the death.

These situations underscore the need for more training and collaboration between law enforcement and mental health professionals. Many communities have adopted dual response programs, where law enforcement initially responds to ensure scene safety before a mental health worker takes over the interaction. While often successful, these programs can be cost-prohibitive and struggle to recruit enough clinical social workers or mental health professionals willing to respond to potentially dangerous situations around the clock.

Most mental health professionals are accustomed to working in controlled, clinical settings – not responding to unkempt public housing units with high crime rates or homeless encampments in the middle of the night. It's estimated that one-third of all unhoused individuals have a severe mental disorder, adding another layer of complexity to these encounters.

Homeless Camp

Too often, I have seen troubling cases involving the intersection of mental illness and homelessness. These individuals are already marginalized by society, and when they're approached by law enforcement in a crisis, the results can be devastating. For many, the police are seen as an enemy, not as a source of help, especially if they've been previously arrested for minor offenses like loitering or public intoxication.

In one case, a homeless veteran suffering from PTSD and substance abuse issues was living under a bridge. Police were called to the scene after receiving complaints about his aggressive behavior toward passersby. The officers, upon arriving, found him wielding a stick and shouting incoherently. Instead of attempting to calm the situation, they immediately went on the offensive, shouting commands and escalating the tension. When he didn't comply, they used a Taser on him. His heart couldn't handle the shock, and he died on the scene.

This death was entirely preventable. Had the officers approached with empathy, offering him the option of connecting with a veteran's service organization or a local shelter, the outcome could have been dramatically different. This is where CIT and mental health crisis training could have made all the difference, helping officers understand that many homeless individuals with mental illness or addiction issues aren't a threat – they're in desperate need of support.

There's a broader systemic issue at play here. Our country has essentially criminalized mental illness, turning jails and prisons into de facto mental health institutions. People who need treatment often end up incarcerated instead, a trend that began in earnest with the deinstitutionalization of psychiatric hospitals in the 1960s and 70s. Without adequate community mental health resources to take their place, law enforcement officers have become the default responders to crises that mental health professionals should handle.

The result is that our jails are overcrowded with people who aren't criminals but, instead, are suffering from untreated mental illness. This puts an enormous strain on our criminal justice system and leaves many individuals without the care they need. According to the National Institute of Corrections, approximately 15% of men and 30% of women incarcerated in jails and prisons suffer from serious mental illness.

These statistics are a call to action for law enforcement, mental health professionals, and policymakers alike. We must invest in diversion programs that keep people with mental illness out of the criminal justice system and in treatment. One promising model is the Law Enforcement Assisted Diversion (LEAD) program, which allows officers to redirect low-level offenders, especially those with substance use or mental health issues, to treatment instead of arresting them.

In cities where LEAD has been implemented, there's been a marked reduction in recidivism rates and a decrease in police use of force. Officers, in turn, have reported feeling more empowered to help people rather than just arrest them. It's a win-win for law enforcement, the individuals in crisis, and the community.

But we can't stop there. Mental health crisis intervention training must become a standard part of police education, from the academy through continuing professional development. Ideally, every law enforcement officer would receive comprehensive training in CIT principles, trauma-informed care, de-escalation techniques, and awareness of implicit bias.

Officers should be taught to recognize the signs of mental illness and understand the nuances of different conditions. For instance, an individual experiencing schizophrenia may respond very differently to police commands than someone having a manic episode. Tailored approaches, rather than one-size-fits-all tactics, are essential in these situations.

There's hope, though. Through sustained advocacy and ongoing research, the field of law enforcement is gradually evolving. We're seeing more departments embracing community policing models that emphasize building trust and cooperation between police and the communities they serve. By fostering these relationships, officers can better anticipate the needs of their communities, including mental health crises, and respond in effective and humane ways.

I've devoted much of my post-academic career to pushing for these changes, working as a consultant and educator. I've had the privilege of training hundreds of officers in CIT principles and as an expert witness in cases involving mental health crisis interventions. Each case reinforces my belief that with proper training and a focus on de-escalation, law enforcement can play a pivotal role in addressing the mental health crisis in America – without resorting to the use of deadly force.

The challenges are immense, but they're not insurmountable. By integrating academic insights, like the ones I explored during my Ph.D. research, with the practical experience of those on the front lines, we can continue to move the needle on this critical issue.

Ultimately, this is about saving lives—both the lives of those in crisis and the officers tasked with protecting and serving them. It's about building a future where mental health crises are treated as public health issues, not criminal matters. And it's about ensuring that law enforcement officers have the tools, training, and support they need to do their jobs safely and compassionately.

Looking back on my career, I am filled with a deep sense of responsibility. I've seen the worst of what can happen when mental illness intersects with law enforcement, but I've also seen the best of what's possible when we approach these situations with care, understanding, and proper training. There's still much work to be done, but I'm hopeful we can make a meaningful difference for future generations through continued advocacy and education.

Chapter 24

The One Percenter's Path

— You've Got To Earn The American Dream! —

Now, I know what you might be thinking. "Roy, you're lucky. You had opportunities that others don't." But I want to challenge that notion. Luck may play a small role but is not the determining factor. The determining factor is what you do with the opportunities you're given, no matter how small they seem.

Throughout my career, from my early days as a Boy Scout to my time as a police chief, military veteran, and beyond, I've come to understand that the American Dream isn't a guarantee. It's not a birthright or a handout. You have to earn it through hard work, perseverance, and a commitment to lifelong learning.

In a nation founded on the principles of equal opportunity and the pursuit of happiness, the concept of the "American Dream" has long captivated the hearts and minds of people worldwide. This dream, the idea that through hard work, determination, and a bit of luck, anyone can achieve prosperity and fulfillment, is the driving force behind the success stories that have come to define the American spirit.

Yet, as inspiring as these tales may be, truly attaining the American Dream is a rare and remarkable feat that places the individual in a rarefied group, a veritable one percent of the population.

The statistics are clear: less than 2% of people in the United States have a Ph.D., less than 1% of the world's population serves in the military and only 1% of law enforcement. I've done all three, not because I'm special or gifted, but because I was willing to do the work when others weren't.

Being a "one percenter" is not about wealth or status but about the extraordinary personal qualities and life choices that elevate an individual above the masses. It's about possessing a relentless work ethic, an insatiable thirst for knowledge, and an unwavering commitment to serving others—qualities I've had the privilege of cultivating throughout my remarkable journey.

From a humble beginning as a C-student in Amherst, Ohio, to earning a Ph.D. and becoming a celebrated expert witness in law enforcement and civil rights, my path has been anything but conventional. I didn't follow the traditional route of academic excellence in high school, nor did I have the advantage of a prestigious military academy or an elite university education. Instead, I carved my way, driven by a deep-seated desire to make a difference in the lives of those around me.

This innate passion for service led me to join the military, not as an officer, but as a young enlistee with a dream of one day becoming a police officer. I understood that the discipline and leadership training I would receive would lay the foundation for a career in law enforcement. This career would ultimately span multiple decades and culminate in roles as a police chief, an emergency management expert, and a national security specialist.

Life rarely gives us a roadmap. Whether you're a five-year-old on your first day of school or a seasoned professional facing a new challenge, the feeling is often the same. You're in uncharted territory, and finding your way is up to you. I've learned that the key to success in these situations isn't having all the answers. It's about having the courage to take that first step, ask for help when needed, and keep moving forward even when unsure.

Throughout this journey, I encountered countless obstacles and setbacks, from the economic challenges of the 1980s that forced me to work multiple part-time police jobs to the emotional and physical demands of my military service in Iraq. Yet, in these moments of adversity, I discovered the true grit and resilience that define the one percenter. I refused to be deterred, letting the present difficulties overshadow my vision for my future.

Whether you're born into adversity or advantage, what matters most is what you do with the hand you're dealt. Take my journey to earning a Ph.D., for instance. I was already working full-time and running my business while pursuing this advanced degree. I had every excuse in the book to say, "I'm too busy" or "I'm too old for this." But I didn't. Instead, I carved out time wherever I could find it. I worked all day, studied late into the night, woke up early to hit the books again, and still made time for my weekend responsibilities.

Was it easy? Absolutely not. Sometimes, I was exhausted and questioned whether it was worth it. But I persevered because I understood that education—formal or through life experiences—is the key to unlocking opportunities and making a real difference.

Indeed, this unwavering determination propelled me to excel in every academy, training program, and leadership role I undertook. This relentless pursuit of excellence and a genuine desire to learn and grow ultimately set me apart from most of my peers.

The same level of dedication applies to those serving in law enforcement and the military. These brave men and women put their lives on the line daily to protect our freedoms and keep our communities safe. They miss birthdays, holidays, and countless family moments because of duty calls. They face dangers that most civilians can scarcely imagine, from hostile combat situations to the invisible scars of post-traumatic stress.

But they do it because they understand that freedom isn't free, and someone must stand ready to defend it. They do it because they know that being part of the one percent means putting the needs of others before your comfort.

The American Dream isn't about being handed success on a silver platter. It's not about luck or waiting for someone else to make it happen. It's about having the freedom to pursue your goals and the opportunity to work towards them. It's about taking responsibility for your success and not waiting for someone else to make it happen. And nowhere is this truer than for those of us who choose to be part of the one percent.

I've seen too many people give up on their dreams because they weren't willing to put in the effort. They look at successful people and say, "They're lucky." But they don't see the years of hard work, the sacrifices, and the relentless pursuit of goals that led to that success.

When building my business, I didn't take a vacation for years. While others were relaxing on weekends, I worked on growing my company. When friends were out socializing, I often studied for my Ph.D. or prepared for my subsequent expert witness testimony. These weren't sacrifices in my mind; they were investments in my future and my ability to serve others more effectively.

I don't regret a single moment of it because all that hard work and those sacrifices paid off, not just in terms of financial success but in the satisfaction of knowing that I've made a difference. I've contributed to making our communities safer and helped shape policies that can save lives. While doing this, I've mentored others and hopefully inspired them to pursue their dreams.

The American Dream for a one percenter isn't about accumulating wealth or achieving fame. Does this mean financial success is off the table? Not at all. Many who follow these service paths find themselves in comfortable positions later in their careers. The crucial difference for us one-percenters is that financial gain is a byproduct of our service, not the primary goal. We've learned that true fulfillment comes not from what you accumulate but from what you contribute.

Too often, many individuals fail to render premium service to others because of their approach. One of the most important lessons I learned in those early days was the value of showing respect—both to the people we serve and to the people we serve with. Service isn't just about upholding the law or maintaining order. It isn't a standalone act of working toward the greater good. It's about treating people with dignity, making them feel loved, and understanding that how you deliver your service is as important as the service itself.

Chief Taylor Dress Uniform

In my years of service, I've come to understand and apply the five universal truths of human interaction. This truth has helped me create a stronger connection and a healthy environment to render my service. They are:

- People feel the need to be respected
- People would instead be asked than be told
- People have a desire to know why
- People prefer to have options over threats
- People want to have a second chance

Service is not just about responding to crises—it's about preparing yourself and others, communicating clearly, and ensuring everyone feels empowered to make the right decisions. It's about being clear, consistent, and always willing to go the extra mile to ensure the people around you understand what to do and why it matters.

To give quality service, you must learn to promote your humanity. The five truths taught me valuable lessons about leadership and the importance of fair, consistent rule application. They showed me how poor leadership and service could create stress and potentially dangerous situations.

It is dangerous if someone tries to be over-efficient and neglects effective human interactions. Therefore, when serving, you must realize who you're dealing with and how to ensure compliance while maintaining your integrity and the respect of those you serve and work with.

So, how does one earn their place among those who render effective service? How does one become a one percenter in the truest sense of the term? It starts with embracing the calling to serve something greater than yourself. It requires a commitment to continuous learning because, in these fields, the moment you stop learning is when you stop being effective.

You also have to develop mental toughness. The path of service is often complicated and thankless. Sometimes, you might meet an individual who has an entitlement mentality, and they may dishonor you with their actions. You must cultivate the resilience to push through challenges, setbacks, and moments of doubt. Integrity must be your north star. In roles of service, your character is your most valuable asset. You must uphold the highest standards of ethics, even when difficult or unpopular.

Building strong relationships is also crucial. No one succeeds alone. Therefore, you must foster connections with mentors, colleagues, and those you serve. These relationships will sustain you through difficult times and amplify your impact. Remember, contributions sometimes come from different teams or departments. Sometimes, the expertise needed lies with individuals or groups outside your immediate circle, and recognizing when to involve them is critical to success.

As you seek opportunities to serve others, always remember that in any field of service, the ability to reach across departments, agencies, or even levels of government service to bring fresh perspectives can make all the difference in solving a problem. When we collaborate beyond our teams, we open the door to creative solutions and ensure every stone is turned.

Perhaps most importantly, you've got to embrace sacrifice. Choosing this path often means giving up personal comfort, time with loved ones, and more accessible career options. But these sacrifices are part of your commitment to service and what sets you apart as a one percenter.

In everything we do, we must learn to know when we are going overboard and endeavor to make time for ourselves and our loved ones. Maintaining this balance is difficult, but it's a necessary part of living a life of service. If we neglect our personal lives, the consequences can ripple out, affecting our families and our ability to serve our community with a clear conscience and open heart. We must remember that our service doesn't end when we clock out for the day—it extends into our homes and relationships, and we must give them the same commitment we give to our roles in public service.

The American Dream is alive and well, but it's not guaranteed. It's a possibility and an opportunity for those willing to work for it. It doesn't matter where you start - what matters is where you're determined to go and how hard you're eager to work to get there. Remember, every expert was once a beginner, and every leader once faced their first day in a new role. What separates those who succeed from those who don't isn't innate talent or luck – it's the willingness to step into the unknown, face challenges, and learn and grow from every experience.

As we wrap up this book, I want to leave you with this thought: You are the author of your own story. Your background and your current circumstances don't define your future. What defines your future are your choices, the effort you put in, and the perseverance you show in the face of obstacles.

If there's one thing I want you to take away from my journey, it's this: Don't wait for opportunities to come to you. Create them. Don't make excuses - make progress. Don't envy others' success - emulate their work ethic.

The American Dream isn't about being in the top one percent financially. It's about being in the top one percent regarding effort, dedication, and impact. It's about living a life of purpose, about using your talents and opportunities to make a difference in the world around you.

So, I challenge you to be a one percenter. Be committed to excellence, dedicated to service, and willing to work hard for your dreams because that defines the American Dream.

Remember, success isn't about where you start but where you're determined to finish. Your journey begins now. Make it count.

Made in United States
Cleveland, OH
04 November 2026

47643613R00142